DISASTER BY CHOICE

DISASTER
BY CHOICE

ilan **kelman**

how **OUR ACTIONS**
turn **NATURAL HAZARDS**
into **CATASTROPHES**

OXFORD
UNIVERSITY PRESS

OXFORD
UNIVERSITY PRESS

Great Clarendon Street, Oxford, OX2 6DP,
United Kingdom

Oxford University Press is a department of the University of Oxford.
It furthers the University's objective of excellence in research, scholarship,
and education by publishing worldwide. Oxford is a registered trade mark of
Oxford University Press in the UK and in certain other countries

Published in the United States of America by Oxford University Press
198 Madison Avenue, New York, NY 10016, United States of America

British Library Cataloguing in Publication Data
Data available

Library of Congress Control Number: 2019945429

ISBN 978–0–19–884134–0

Printed and bound in Great Britain by
Clays Ltd, Elcograf S.p.A.

To everyone who has suffered needlessly in disaster
To us all

PREFACE

Disasters are not natural. We—humanity and society—create them and we can choose to prevent them. That is the main message of this book.

Stating that natural disasters do not exist because humans cause disasters seems insanely provocative. We witness nature ravaging our lives all the time: from a city underwater after a hurricane to rows of smouldering houses after a wildfire to the dust rising from the ruins after an earthquake. How could we withstand the 400 kilometre per hour winds of a tornado, faster than Japan's bullet trains, or the 1,200°C temperature of lava, hotter than many potters' kilns? How would we feel if an 'expert' lectured to us that it was not nature's fault, as we sifted through the few photos salvaged from the pile of debris which was once our home and our life?

Yet even when we cannot keep our infrastructure standing, we can stop people dying, we can protect our most valuable possessions, and we can learn to deal with devastation. The disaster lies not in the forces unleashed by nature, but in the deaths and injuries, the loss of irreplaceable homes and livelihoods, and the failure to support affected people, so that a short-term interruption becomes a long-term recovery nightmare.

The tornado, the earthquake, the tsunami are not to blame. They are manifestations of nature which have occurred countless times over the aeons of Earth's history. The disaster consists of our inability to deal with them as part of nature. We have the knowledge, ability, technology, and resources to build houses which are not ripped apart by 400 kilometre per hour winds. If we choose to, we can create a culture with warning and safe sheltering. Lava at 1200°C and a tsunami higher than our building are harder to ride out. But we can shun places likely to be hit by them or we can create a culture which understands and accepts periodic destruction, again with warning and safe evacuation, to permit swift rebuilding afterwards. The baseline is that we have options regarding where we live, how we build, and how we get ourselves ready for living with nature. Many of the choices we make currently permit death and devastation. They create the conditions for disasters.

Nature does not choose, but we do. We can choose to avoid disasters, and that means disasters are not natural.

ACKNOWLEDGEMENTS

This book was made possible by family, friends, my agent and her team, my editor/publisher and her team, and reviewers who proofread and fact-checked. Researchers, policymakers, practitioners, and many others provided the ideas and conclusions in this book long before I had started writing on this topic and they continue to inspire and direct me. They deserve all credit for the originality, innovation, and vision of this book's content, while all errors are mine alone. No advice or recommendations are implied by any content in this book.

CONTENTS

CHAPTER 1

AN ISLAND SHATTERED

An Earthquake Strikes

Another hot, humid, dusty day in the streets of a Caribbean capital. Flies crowd the markets and aged motorbikes and cars bounce through the potholes, coughing emissions and dodging pedestrians. Azure waves lap against the docks while street dogs and pigs scrounge for morsels in the trash heaps dotting the hill neighbourhoods. Just the typical sights, sounds, odours, and haze of life playing out as it has thousands of times before, with the shimmering heat starting to dissipate as the sun droops towards evening.

Today, though, is far from typical. It is 4.53 p.m. on Tuesday 12th January 2010 in Port-au-Prince, Haiti. Twenty-five kilometres west-south-west of the city, just thirteen kilometres below the surface, the earth jolts, fracturing rock, heaving soil, and radiating waves of motion with an intensity technically described as moment magnitude 7.0. This is high on the scale with which earthquakes are measured; just one or two earthquakes at this level or stronger occur in any month somewhere in the world, although often much deeper below the surface than Haiti experiences on this day.

The earthquake's first set of waves, the primary waves, travel twenty times faster than most passenger jets, giving premonitions of the coming destruction. Port-au-Prince has the misfortune of being so close to the rupture's centre that most people have only seconds before the secondary set of waves, the shear waves travelling half as fast, arrive. The devastation begins.

The arrondissement (district) of Léogâne, almost on the earthquake's epicentre, witnesses over 80 per cent of its buildings badly damaged or destroyed. Many rural areas lose the little infrastructure they have, with dirt roads cut by landslips and schools entombing pupils and staff. Around the capital, thirty kilometres away, the shaking lasts for between thirty and sixty seconds. Thousands of buildings crumble, especially the shoddily built, ramshackle dwellings inhabited by people with no choice but to live there. Some buildings collapse in on themselves or tip over. Others slide down the hills into the ravines. The stories of the people who live in these homes are rarely told: people who scrape by day to day, ambling along the unpaved roads, living without running water, electricity, or sewage systems, and then dying in the ruins of their small, dilapidated shelter.

As one lives poor, one dies poor. The immediate aftermath could not bring ambulances and fire trucks wailing through the streets, with hard-hatted rescuers ready to haul out unconscious survivors, stabilize them right there, and then whisk them away to advanced life support in hospitals. Instead, roads are impassable to cars, vans, and trucks—as many were before the earthquake, with these vehicles stymied by the steep slopes, the inadequate infrastructure, and the difficulty of driving after dark. Rescue and medical services do not exist to respond, and this too is a situation long predating the earthquake. As the sun sets, the survivors wonder what to do about water, food, toilets, and sleeping—questions many of them asked every day.

Except that today, the earthquake has happened. The first task is finding family and neighbours under the rubble. Throughout the dusty area, people scramble onto and into partially standing structures, despite the danger of aftershocks. They dig with shovels, makeshift tools, and their bare hands, carrying blood-covered casualties outside.

The destruction seems arbitrary. A pile of masonry which was once a house sits sandwiched between two others that still stand; a collapsed roof is surrounded by others that remain intact. Other blocks, which were crowded neighbourhoods earlier in the day, now resemble derelict construction sites. As the dead are recovered and the injured rescued, makeshift medical centres—never with enough personnel or gauze or room or antiseptic—shelter child and adult amputees. Soon, tent cities spring up proffering small rooms under flimsy sheets for extended families that have lost half their members.

The quagmire of poverty perseveres. Before 12th January, how could Haitians in the informal settlements begin to think of earthquakes? Even where their abodes were not constructed from masonry, sometimes making them safer during earthquakes, daily life meant struggling for food and water, navigating the open sewage and the violence, and being denied education and healthcare. Not much changes after 12th January, except that their rickety residences exist no more.

Higher-profile structures of the well-to-do fare no better and the losses are just as calamitous, but they garner international attention and international rescue resources. Parts of the Presidential Palace crumble, leaving it like a smashed and smeared wedding cake. One section of the country's United Nations (UN) headquarters pancakes, six storeys becoming one. UN employees die, 102 in all, the largest single-day loss of life in the organization's sixty-four-year history and a tragedy involving staff from every inhabited continent. The dead include the top UN official for Haiti, his deputy, and the acting police commissioner—all well deserving of the praise and memorials they received, as are the tens of thousands of ordinary Haitians who were provided with neither.

Many other UN employees and Haitians survive, only to reach home to discover that their spouses and children did not. A Danish man who had been working in the UN building is rescued five days later. He describes how he heard others trapped but alive, their noises ceasing thirty-six to forty-eight hours before he was pulled out. Dust clouds lift above the luxury seven-storey Hotel Montana, renowned among foreigners and once hosting dignitaries, celebrities, and country leaders. Over four dozen bodies are eventually recovered, including staff, tourists, and business travellers.

As the earthquake runs its course, at least 150,000 people and possibly up to double this number, the vast majority of them poor Haitians, are dead or dying. Disaster casualties are notoriously hard to tally, especially when the number of people living in an affected place is not known. The true death toll, and the true toll of suffering, in southern Haiti, can never be known.

The Haitian disaster mobilized the world. Aid soon poured in, along with journalists. Taking over from locals using their bare hands and basic tools to dig for survivors, international rescue teams crawled through collapsed structures time and again.

Many tearful reunions made it worthwhile, but the joy was marred by the growing piles of corpses. Locating all the bodies took weeks. When it was deemed that no one remained underneath a wrecked building or that they were not recoverable, a major logistics operation cleared and dumped the masses of rubble.

Governments and organizations pledged around US$13 billion of aid and delivered perhaps half. Remittances and individual donations are harder to track and they provided support to people who had lost everything except their own lives. The people's continuing desperation did not stop the political shenanigans. Days after the earthquake, the US military took over airport operations at Port-au-Prince, sparking a backlash from countries and agencies whose aircraft were unable to land. As the political fights brewed, Haitians were left needing the basics.

Flimsy tents were erected on mud-prone slopes only to be blown away by moderate winds. Privacy, security, dignity, and safety were not always significant considerations in setting up the temporary settlements, leading to continual rape and assault with little recourse for catching and punishing the perpetrators. Even allowing for the fact that attacks are under-reported because many fear that they will be stigmatized or abused for documenting what happened to them, by the end of a year after the earthquake, at least one-tenth of households in the temporary settlements had reported that a member had gone through some form of sexual assault.

As the post-earthquake months dragged on without promised housing being built, tens of thousands of Haitians were forcibly evicted from these temporary settlements. In many cases, they went from squalor to nothing. Once again, so many people were left without safety, without jobs, without healthcare, and without schools for their children, despite the world's attention and the promises to 'build back better'.

The troubles did not stop there, as UN soldiers tasked with helping post-earthquake reconstruction made the situation even worse. On 21st October 2010, the Haitian government declared an outbreak of cholera, a disease not seen in the country for over a century, and over 10,000 people have died from it since. The UN soldiers had introduced cholera, which spread swiftly due to the extremely poor quality of water, sanitation, and hygiene across Haiti, not helped by the earthquake disaster.

The UN first tried to avoid responsibility for the outbreak, but then commissioned an independent report which concluded in May 2011 that the methods for handling and disposing of the UN soldiers' human waste were not sufficient to prevent cholera contamination. Claims for compensation from the UN by those affected by cholera were met with intransigence up to the level of Secretary-General. In February 2013, a formal UN statement effectively denied the need to provide compensation by not accepting legal responsibility for the cholera outbreak due to diplomatic immunity.[1]

It took until August 2016 for the UN to admit formally the role its troops played in bringing cholera to the country and it was December of that year when UN Secretary-General Ban Ki-moon, in his last month holding his position, finally issued an apology.[2] In between, in October 2016, the UN offered 'material assistance' (which some take as a code word for 'compensation') to those affected by cholera as part of a US$400 million initiative, which also involved eradicating the bacteria and improving Haiti's water and sanitation infrastructure. Without a specific funding or allocation plan for this initiative, it was unsurprising that, over two years later, little had happened and cholera still plagues Haitians.

From flimsy shelters to disease to sexual violence, how could so much go so disastrously wrong during the relief, recovery, and reconstruction? There is no lack of experience in humanitarian aid and no shortage of guidelines and manuals gracing the shelves. All the heartaches identified in post-earthquake Haiti had multiple precedents registered in reports from past disasters.

More to the point, why did the world have to wait for the earthquake to mobilize so much help for Haiti? As with humanitarian aid, there is plenty of experience side by side with guidelines, manuals, and texts on building to withstand earthquakes and other hazards—and developing a society which can deal with an earthquake and other hazards. Little of this accumulated insight had been applied to Haiti. The promised US$13 billion, or even half of that, would have gone a long way towards preventing the disaster in the first place.

Before and After the Earthquake

It is hard to predict exactly when earthquakes will hit, as there are few reliable warning signs beforehand. In the long term, statistical analysis can provide some indication of the time frame within which a geological fault will move. For now, so many assumptions are needed, and the ways in which a fault can shift are so varied, that confidence in the predictions is low. In the short term, hours or days before a tremor, a variety of proposed indicators has been examined, from radon gas to groundwater levels and from animal behaviour to electricity in the air. No single sign has proved itself sufficiently to be accepted. This doesn't, of course, stop some individuals popping up after a major earthquake claiming they foresaw it, neglecting all the previous times they predicted an earthquake which did not happen, and so making scientists rightly leery about such claims.

Despite not knowing when an earthquake will occur, many major tremors can generally be forecast by location, in terms of mapping known faults, most notably along the boundaries of tectonic plates. This method is not 100 per cent reliable and earthquakes still occur far from major tectonic plate boundaries, as witnessed on 25th December 1989 when northern Québec, in the middle of a tectonic plate, experienced surface ruptures during the Ungava earthquake. Even at known tectonic plate boundaries, previously unknown faults can slip, which is what occurred in Northridge, California on 17th January 1994, with damage killing about fifty-eight people. We certainly know that Dushanbe, Istanbul, Jakarta, Kingston (Jamaica), Mexico City, Quito, San Francisco, Skopje, Tehran, Tokyo, Vienna, and Wellington among many others will be rattled at some point. Port-au-Prince was also on this list.

Knowledge of Haitian earthquakes extends back centuries. In 1842, a quake about ten times as powerful as that of 2010 rocked the country's north coast. Hesketh Prichard, a British explorer and then a First World War sniper, referred to it in a scientific article of 1900, describing his journey across Haiti.[3] Another research piece from 1912 mentions 1842 alongside the major damage around Port-au-Prince from shakings in 1751 and 1770.[4]

Yet despite this knowledge of seismicity, little was done. Why was the infrastructure in the capital city so poorly constructed? Why were so many people poor, leaving them with no choice but to live in these buildings without hope of improving them? Why did even the affluent parties, from the country's president to the UN to the developers of Hotel Montana, not enact basic earthquake safety principles? These questions were being asked in 2010: a meeting on tackling disasters in Haiti, highlighting seismic safety, was concluding on 12th January when the earthquake rumbled through.

The overwhelming inequities, underdevelopment, and marginalization precluded a quick fix. Haiti, as a country, is not especially poor or under-resourced, but the scale of inequality is horrifying. Centuries in the making, all these problems could not be easily solved. It takes time to put up tens of thousands of buildings which lasted barely a minute on 12th January. It takes time to create a city rife with informal settlements, without basic services, and lacking planning regulations, building codes, and institutions to monitor and enforce such laws. It takes time to produce a culture of day-to-day bustle across exposed electric wires, through haphazard doorways, and around informal structures.

For Haiti in 2010, this time period might have been precisely 206 years. Before 1804, France had been the ruling colonial power, exploiting slaves to plant and harvest coffee, sugar cane, and tobacco. In treating human beings as commoditized objects, France's costs were low and profits were high, a wealth built on the back of inhumanity. At the time, a colonizer would not care much about saving the lives of the colonized, especially since slaves could easily be imported from elsewhere.

In 1791, in the wake of the American War of Independence (1775–83) and the French Revolution of 1789, Haitian slaves rebelled. Twelve years later, as the Napoleonic Wars gripped Europe, the Haitians won. A year after that, on 1st January 1804,

Haiti declared independence, so that the first free Caribbean state was born, only to remain everywhere in chains.

Colonial powers refused to accept this freedom. France exacted a heavy toll through a demand for reparations, finally paid off by Haiti in 1947. The USA, with its own colonial experience apparently forgotten, joined in the exploitation. In 1868, American President Andrew Johnson suggested annexing the island, but it never happened. US warships were a common sight around Hispaniola until in 1914 US President Woodrow Wilson sent in the marines to move the foreign cash reserves of Haiti to New York. In 1915, the marines arrived again as an occupation force, staying until 1934.

This pattern of control persisted over the next decades through Haitian leaders. François ('Papa Doc') Duvalier retained an iron grip over the country from 1957 to 1971 after which his son Jean-Claude ('Baby Doc') Duvalier continued in the same vein. They subjugated, tormented, and pillaged Haiti as much as France and the USA had done before them. Although the former colonial powers were not wholly enamoured with the Duvalier regimes, they more often than not took advantage of the leaders' absolute power while the Haitian people suffered. Baby Doc fled to France during a popular uprising in 1986, paving the way for elections and coups in Haiti.

Jean-Bertrand Aristide became the on-again, off-again elected president of Haiti. A convoluted American foreign policy remained uncertain exactly where the White House and American troops stood with respect to Port-au-Prince. In 2004, the UN took over and was working at reconstructing the country with evidence of some progress.

After a steady increase between 2004 and 2008, Haiti's population growth rate dipped slightly in 2009. Infant, child, and maternal death rates continued their steady decline, each dropping by more than 10 per cent from 2004 to 2009. Numbers and rates of undernourishment decreased during the same time period.

With over half the population under the age of 20, opportunities for education and jobs persisted as a major challenge. Riots over food prices in 2008 precipitated the firing of the prime minister. As Haitians had done for decades before, thousands fled in derelict boats, hoping to reach the USA, the Turks and Caicos Islands, or the Bahamas. If they did arrive and could find work, their remittances provided a lifeline for those staying at home. Commonly, the US Coast Guard intercepted and repatriated those in boats, leaving them to try again another day. Many more simply disappeared at sea, slipping below the waves or being devoured by sharks.

This roller-coaster progress characterized Haiti, as it had done since independence, when the earthquake battered over two centuries of social and infrastructural neglect. And earthquakes are not the only hazard facing Haiti. On 14th September 2004, tropical storm Jeanne formed in the ocean near several Caribbean islands. The system tracked west-north-west, becoming a hurricane on 16th September and skipping along the Dominican Republic's north shore as a tropical storm before unleashing its rainfall across northern Haiti.

The city of Gonaïves was worst affected, suffering more than 2,800 of Haiti's 3,000 Jeanne-related fatalities. The same vulnerabilities which led to the 2010 earthquake's devastation created the Jeanne disaster. Lack of opportunities, gross inequities, oppressive dictatorships, and centuries of exploitation by the outside world made people vulnerable.

Part of this equation was environmental degradation, including decades of deforestation. Denuded of trees, the hills sent the rainfall sluicing into low-lying areas. Mud, floods, and landslides marked Jeanne's passage, killing people and blocking roads delivering post-disaster aid. The same was true outside of hurricane season. Four months before Jeanne, along the Haiti–Dominican Republic border, flash flooding from intense rainfall killed more than 1,000 people in Haiti and over 400 in the Dominican Republic.

It would be easy to blame the rainfall for leading to deadly floods and landslides. It would be straightforward to pontificate that Haitians made the decision to cut down the trees. Doing so ignores our understanding of the nature of these unnatural disasters. The real disaster is revealed by the questions which we must raise and answer. Why did people feel they had no choice but to cut down trees? Why was infrastructure so poor that it could not withstand rainfall? Fundamentally, why did people not have the resources, knowledge, options, abilities, and opportunities to prepare for and deal with a storm? The answers to these questions are the same as for the 2010 earthquake.

Then, on 29th September 2016, a tropical storm formed just west of Barbados. As hurricane Matthew, it briefly reached Category 5, the most powerful, on 1st October before shifting between Categories 3 and 4 during its march northwards. Jamaica, Cuba, Haiti, and other countries in the area issued warnings. Many people evacuated locations in Jamaica as Cuba's well-tested civil defence went on standby. But Haiti took the brunt of the storm when it tracked east, cutting across the country from 4th to 5th October. Jamaica escaped a direct hit, with no reported deaths. Cuba and the Dominican Republic each listed four fatalities. Haiti's toll was at least 500 killed.

And there was the continuing cholera epidemic too. Following Jeanne and the May 2004 flooding, at least there had been no need to worry about the spectre of cholera. Society had taken action to deal with cholera long before 2004. It was the same during the 2008 hurricane season. Four storms in a row—Fay, Gustav, Hanna, and Ike, representing two tropical storms and two hurricanes—ripped through different parts of Haiti killing hundreds. No matter how much or where the rain fell, cholera had never been a concern.

By the time hurricane Matthew appeared on the map, Haiti had already reported 29,000 cholera cases for 2016. The disease's death toll around the country since it was introduced after the earthquake in 2010 already matched, if not exceeded, the total number of storm-related deaths within the same time period. As Matthew moved on and the floodwaters subsided, Haiti's cholera cases grew. The humanitarian response was duty-bound to involve cholera prevention, treatment, and vaccination. Human decisions following the 2010 earthquake had created the hazard of cholera for Haiti, which existing vulnerabilities turned into a continuing disaster. This disaster was illuminated by another hazard, a hurricane.

No shaking of the earth, no downpour from the clouds, and no wind from a storm created cholera in Haiti or vulnerability to the disease. The disaster is not natural. The virus was introduced by people, and it continues to grip the country because of human failures.

Two months after Matthew, an international group of doctors and researchers advising the Minister of Health and Population of Haiti estimated that cholera transmission in Haiti could be stopped within five years for around US$66 million.[5] Considering cholera's cost of thousands of deaths and hundreds of thousands falling ill, as well as what might transpire should the disease cross the border into the Dominican Republic, this is a bargain. Thus far, no one has given the resources needed to achieve this goal, even though we know that every year, Haiti has the potential for a hurricane, earthquake, or tsunami which would entail the emergency import of cholera treatment and vaccination equipment.

Such decisions are not one-offs. They are systematic and continual, ensuring that the cholera burden on the Haitian population endures. They exemplify a long-term attitude of, in effect, allowing the perpetuation of a long-term problem inside Haiti which was introduced from outside.

And so the disasters continue. They parallel exactly the perpetuation of the long-term problem inside Haiti of entrenched disaster vulnerabilities which were introduced mainly from outside the country. The systematic and continual decisions to oppress most of the Haitian people, to snatch resources from the country, and to sustain the abject poverty all conspire to create the vulnerabilities which in turn create the disasters. The 2010 earthquake exposed the centuries of neglect, brutality, and vulnerabilities foisted on the Haitians who could least afford to challenge their locked-in position.[6] The 2016 hurricane exposed the years of neglect, brutality, and vulnerabilities foisted on the Haitians who are least able to avoid cholera by their own means.

This is the disaster. The disaster is these long-term processes, over years and centuries, not the short-term events, over seconds (earthquakes), minutes/hours (tsunamis), and days (hurricanes). The process of unrolling disaster is based on the long-term choices of people who have the power, resources, knowledge, and abilities to make essential and intrinsic changes—but apparently not the wisdom, will, or principles to do so. A disaster is not an event and is not the fault of nature. A disaster is a process manufactured and implemented by people and their choices.

How could an oppressed, marginalized, overexploited country forced to remain underdeveloped with the people in poverty ever deal with nature's extremes? Hurricane Matthew provided at least three days of preparation for Haiti. The entire Caribbean knows that any date between June and November (and sometimes outside these months) could bring a storm roaring through. The 2010 earthquake provided barely seconds of preparation time, but Haiti nevertheless knew that it could shake at any time.

Disasters such as those hitting Haiti hit the headlines and capture our attention. They are nothing new, having always happened throughout human history. Did disasters exist before human history? This question is not easy to answer because a disaster is described through effects on humans and society. Neither the earthquake of 2010 nor the rainfall of 2004, 2008, and 2016 would have mattered if they had not killed and injured people, disrupted routines, and damaged infrastructure. It is hard to have a disaster without humanity. Yet nature still produced the earthquake and storms.

Disasters not involving nature—such as chemical explosions, riots, and terrorism—are clearly not natural. When an environmental component—such as earthquakes and hurricanes—is involved, then disasters seem to be caused by the environment and are blamed on nature. Then what exactly is wrong with the phrase 'natural disaster'?

The definition of the term 'disaster' relates to its impacts on humans. At the basic level, trawling through hundreds of pages of academic writing on the definition, dozens of professional manuals, and several dictionaries, a reasonable definition is 'a situation requiring outside support for coping'. Something happens, we cannot deal with it, and we ask for help. This concept works at the individual level and at the international level, matching UN glossaries, researchers' viewpoints, emergency services' interests, and dictionaries.

These seven words display vagueness—how should 'situation', 'support', and 'coping' be interpreted?—but vagueness can rarely be avoided. The principal power of these words is that they are understandable, somewhat intuitive, and work across many (although certainly not all) languages and cultures. The key is that disasters are defined by their societal impacts, not by the degree or scope of any influence from nature.

We also need to consider why those affected by nature cannot cope with it. The reason lies beyond the natural environment: vulnerability of people, places, infrastructure, and communities. Vulnerability means that people do not have the resources, knowledge, or choices available to stop nature from harming them.

Haiti's earthquake and storms were natural, but its disasters were not. They arose from individuals and society. But this is all for Haiti. Does the same pattern hold elsewhere?

CHAPTER 2

NATURE'S HAZARDS

Change is Natural

Our natural environment changes, as it has done for billions of years. Species evolve, continents drift, and sea levels fluctuate over time periods of hundreds, thousands, and millions of years. Some volcanoes wait millennia or longer between eruptions, as magma from deep underground slowly wends its way to the earth's surface. In the meantime, people build. The last known eruption of Germany's Laacher See volcano near Bonn was nearly 13,000 years ago, when comparatively few people were in the area. Scientists calculate that a similar eruption today would affect over two million people and the damage to housing alone would cost between 18 billion and 27 billion euros.[1]

As the ages wax and wane, the shape of the earth's orbit around the sun varies. So does the direction in which the earth's axis tilts compared to the sun. These variations affect seasonal extremes and lead to ice ages advancing and retreating.

Over human lifetimes, across decades, the climate witnesses further changes. The North Atlantic Ocean, the Pacific Ocean, and the Indian Ocean each dance through slow oscillations of their properties such as temperature distributions and air pressure, affecting regional and global climates. On land, over decades, forests mature, expand, and shrink while, over slightly longer time frames, soils form underneath the canopies and across grasslands. Land, oceans, and the atmosphere have been shifting over decades from human-induced climate change. As the air warms by fractions of a degree Celsius each year, average sea levels creep up by millimetres per annum and ocean acidity increases. Many glaciers and ice sheets have been steadily shrinking while others expand with increased snowfall due to climate change. In places, erosion of topsoil, coasts, and river banks is easily measured over months and years. Some coastlines around England annually retreat more than a metre. Expanding coastlines elsewhere include some salt marshes of New York, which continue to rise a few millimetres above sea level each year, along with much of Norway's shoreline.

These changes are gradual. All the same, infrastructure which does not adjust to them becomes damaged, for example when the subsidence or rising of land causes buildings to crack and fall apart. Sometimes the drama even plays out on live TV. In June 1993, the four-star Holbeck Hall Hotel built in 1879 near Scarborough, England was videoed tumbling over a cliff as a landslide glided into the sea over several hours.

Hurricanes, wildfires, temperature extremes, and river floods can last days. Tsunamis and thunderstorms rarely endure longer than hours. Earthquakes typically complete their shaking in seconds. Consequent damage makes these creeping or rapid changes hazardous to our infrastructure and disrupts our lives. When we have not planned or prepared for it, nature's hazards over any time period lead to damage and losses, to life, livelihood, and infrastructure—in effect, a disaster. The key is 'when we have not planned or prepared for it'. In none of these cases is nature intent on being malicious.

Walt Disney's 1942 movie *Bambi* reflects our view of nature. Who could forget, it is suggested, the forest fire that endangered Bambi's life?[2] He and his father bounded through the woodland with sparks showering around them and trees in fiery explosions barring their way. Yes, 'Man' had entered the forest and 'his' campfire had set the woods aflame, forcing the animals to gather in the safety of a lake island.

Characters from Bambi soon fronted forest fire prevention campaigns. The bear Smokey supplanted them in 1944, morphing into Smokey the Bear in a 1952 song. In folksy style, 'the fire preventin' bear' lamented 'what you'll be missing' if all trees 'went up in smoke'. Thus was born the phrase 'Only you can prevent forest fires'. All fire is bad, so stop all fire. To be fair, Bambi and his friends were pursued by unnatural flames, from human carelessness. In any case, forest fire prevention in the USA long precedes Disney fans' favourite fawn. Does Bambi really bear the blame for American post-Second World War policies which aimed to suppress what are now called wildfires?

After all (spoiler alert), a hunter felled Bambi's mother. Bambi's plaintive cries as he fruitlessly searches for her in the falling snow under darkening skies have left the USA's hunting and gun cultures intact. The deer-child's single teardrop did not even impede hunters from expressing offence at their portrayal in the movie. Somehow, bullets against nature are acceptable, it seems, while fire is not. But preventing fire from human mistakes, or deliberate setting, should not shift to obstructing all wildfire. Nature is full of change, and fire is part of nature.

Wildfires are nevertheless terrifying. They can advance at more than ten kilometres per hour in forests, faster than most people can walk, or double that speed across non-forested land. The air preceding a wildfire can exceed 800°C as flames leap dozens of metres upwards, cresting the tallest tree crowns. Sparks and debris drift along or are fanned by the wind, igniting land and property far from the main front of the fire. Initial triggers might be lightning, sparks from power lines or motorcycle engines, vehicle fires, cigarette butts, neglected campfires, or, worst of all, arson. It is unsurprising that significant efforts are put into controlling and stopping wildfires.

The morning of 30th January 2009 dawned hot and dry over the state of Victoria in Australia. Melbourne recorded a peak of 45.1°C, one of the highest formally measured temperatures in the city. Similar conditions had persisted for the previous two months, leaving vegetation around the state parched. As the lack of humidity and high temperatures endured over the next week, the premier of Victoria warned on Friday 6th February that the following day would bring Victoria's worst-ever recorded conditions for vegetation fires.

Australia has good reason to fear the flames. The incongruously named Ash Wednesday fires of 16th February 1983 killed seventy-five people across two states in twelve hours. Tasmania lost sixty-two people to the Black Tuesday fires of 7th February 1967. Forty-two years later to the day, their Victorian compatriots waited to see whether or not their fire warning would anticipate a disaster.

Saturday 7th February 2009 brought near-hurricane-force dry winds sweeping across Victoria. At around 11:45 a.m., a power line failed, partly due to an incorrect installation which had been missed during an inspection the previous year. Electric arcing from the failure ignited the vegetation, which was, in the premier's words, 'tinder-dry'. A few minutes later, an observer in a fire tower reported smoke. Fire crews were alerted within three minutes and were on-site before noon. But the fire was already out of control. It jumped roads and advanced through the bush along multiple paths. Sparks fanned by gusts ignited new fires up to forty kilometres ahead of the main front. Flames leapt over fifty metres high.

The rapid, erratic spread and the continuing, multiple ignitions—including from arson—coupled with the wind's change of direction earlier that evening left many people little time to prepare or flee. Several firefighters found themselves caught in the fire but survived. Risking their lives, they saved hundreds of others. Sadly, the flames trapped dozens more. By the time this bushfire had completed its run, 119 people had perished and 232 were injured, some with horrific burns.

Overall, 7th February, or Black Saturday, realized an even higher toll. All the day's bushfires around Victoria together killed 173 people and injured 414. Over one million animals were killed and 3,500 buildings destroyed. It is Australia's worst bushfire disaster so far.[3]

How could such vulnerability to a known hazard arise?

Indigenous Australians managed fires for tens of thousands of years. They set controlled blazes to alter the environment for maintaining tracks, trapping animals, and avoiding the build-up of burnable fuel which could lead to large conflagrations. Over time, indigenous practices adapted the ecosystems to support plant species which could survive low-intensity bushfires, actually using fire to propagate. Fire was part of land use and land management, integrated into human needs among other environmental adjustments, although we do not really know how many fire disasters the indigenous Australians might have caused nor how many of them perished in the flames.

Europeans imported and imposed a different perspective of bushfires. Flames were presumed always to be dangerous and damaging, so they were suppressed and fought. As settlements expanded into the bush, fires indeed became highly destructive and lethal, reinforcing the combat mode.

The same is true across North America. Wildfire is part of the ecosystem and it is a needed ecosystem process. Californian and Coloradan forests, meadows, and scrubland would not exist today without occasional burning. Suburbia has sprawled into these areas of vegetation and their fires. How could we help ourselves and nature by living with natural fire rather than harming both by manipulating it?

The theory is that preventing wildfires delays the inevitable. Ecosystems expecting frequent, lower-intensity fires might have trouble with the changed regime of less regular flames. Leaves, plant litter, and dead trees build up, providing large swathes of combustible fuel during dry spells. Then, a rare fire rages as a high-intensity, hard-to-control inferno destroying plants, animals, people, and infrastructure.

As we entered the twenty-first century, debates on vegetated lands and wildfire management continued from North America to Australia. Prescribed burns to reduce fuel loads appeared to reduce the intensity of fires, but remained controversial, particularly when properties could be at risk. Some evidence countered the notion that human fire prevention strategies undermine natural fire cycles and lead to worse fires. Survival strategies for people in fire's way came under question, as did the role of climate change in affecting heat, humidity, and winds.

What persisted unquestionably was urban expansion into fire-prone locations. From Calgary to Canberra, dwellers at the city–wildland interface enjoy the quality-of-life benefits of leafy green surroundings, less air pollution, and nature-based activities right beside them. They sit in the middle of areas which are not only flammable but which also require periodic burning for ecosystem health.

Since periodic burning is not healthy for houses or people, a balance is still sought between healthy ecosystems with wildfires and not placing people and properties at risk. Much of the advice must centre around the assumption that fires will happen. Sparks, embers, and flames must eventually envelop properties that infringe on locales which were previously used to being burned.

Warning, preparedness, and responsiveness are essential. We can plan to stay and defend our homes, but extensive preparation and care are needed, as well as being psychologically and physically ready. One small mistake could end our lives. Other strategies are keeping surrounding land clear of burnable vegetation, applying proper landscaping, being attuned to environmental conditions and information sources, and using fire-resistant roofs, walls, doors, and windows. To survive an evacuation, strategies incorporate practising and implementing an escape plan, protecting irreplaceable valuables, having insurance, and being psychologically and financially ready to rebuild. And, especially, leaving long before the flames approach. No strategy is foolproof. All reduce vulnerability to some degree, especially keeping options open and deciding quickly once a threat is palpable—and preferably before the situation is urgent. No matter what, planning and preparedness must begin long in advance and must never stop.

At two and a half kilometres above sea level, people used to living close to sea level quickly end up short of breath when they exercise. Those who grow up or have lived for a long time in Colorado's Rocky Mountains, a more than two-hour jet flight from the nearest salty shorelines, have adjusted. They rarely notice the reduced atmospheric pressure, which means less oxygen, revelling in the alpine air, culture, wildlife, and woodlands. With the forests come all parts of forest ecosystems, including wildfire, which is never too distant.

Houses in the woods surrounding Nederland, Colorado sit about as high as one can live in this area. Some residents were born in the town and never left. Others are transplanted, mainly from around the USA. Everyone deals with the quirks of Nederland, noted for its annual March festival called Frozen Dead Guy Days. Centred around a cryogenically frozen Norwegian who was shipped out to Nederland by his family, the festivities include coffin races, a polar plunge into freezing water, and human 'foosball' (which is life-size table soccer, so people play the game themselves rather than controlling plastic effigies attached to bars).

Nederland's foibles are not just cultural, but also arise from nature. To live there, you need to learn about the blizzards and winds buffeting the canyons and about the moose which can suddenly step out in front of your vehicle as you drive. You need to be aware of the squelching downpours that cause lethal flash floods, especially because a cloudburst upstream means that the flood can swiftly sweep through a sunny location. In any case, the scorching summer days demand continual hydration, as the lack of humidity sucks you dry. Among all these hazards, you must certainly learn about wildfires—and what to do about them.

The 9th of July 2016 was yet another toasty, dry summer day on the lee side of the mountains and foothills around Nederland and down the canyon to the city of Boulder, which extends out onto the plains. Residents and visitors expect such weather. The myth of 300 sunshine days per year and the vast tracts of accessible, scenic parkland beckon nature lovers from ambling tourists to serious climbers. As we stroll along, we may see a deer bounding through the grassland at sunset or prairie dogs perched on their hind legs twitching their noses and waggling their tails at us. Bouldering and cycling routes entice recreationalists who cool off afterwards by tubing down the creek chilled by snow and glacier melt. Birdwatching binoculars train on the varied raptors while their owners hope a bear does not ramble into view.

Some people head out for a few hours for a relaxing picnic by emerald lakes. Others camp for a few days, trekking deep into the backwoods and scrambling over the scree. Two men and a woman from Alabama chose July 2016 to camp around the woods of Nederland. On 8th July, the two men did not properly extinguish their campfire. Twenty-four hours later, the flames lit up the forest in what became known as the Cold Springs Fire. It burned a swathe through the tinder of trees, forcing nearly 2,000 people to evacuate, killing numerous animals, and destroying eight homes along with several other buildings. Fortunately, no one died.

The campfire trio were soon arrested, and trespassing, arson, and other charges followed. The two men had been responsible for the campfire, so the woman plea bargained, receiving community service and probation. The men pleaded guilty and, after four months in jail, were sentenced to a programme permitting them to be employed provided that they return to prison after work. It will take them the rest of their lives to pay for damages awarded against them.

As this drama was unfolding in a Boulder District courtroom, some residents in the fire zone were cleaning up from the blaze—but they were not rebuilding their properties. The flames had swept through their land sparing the houses. Not 'miraculously sparing the houses', because there was no miracle. Foresight, initiative, planning, and actions had saved these homeowners from ruin.

Wildfire Partners is a local and state government funded programme inspiring and supporting residents to implement measures countering wildfire damage to their properties. They provide assessments, detailed advice, progress checks, and occasionally some financial assistance to enact recommendations. In the Cold Springs Fire of 2016, eight houses participating in the Wildfire Partners programme were in the burned area. All survived and were habitable immediately afterwards.

To avoid embers drifting inside, gaps and holes in walls and roofs must be covered or closed while skylights and solar panels are kept clear of debris and litter. Vents, doors, and windows can all use improved materials and construction to reduce the chance of the building catching alight. Fences, porches, and decks require non-combustible material and should be free of other combustibles on or under them. Woodpiles are placed away from the house on a non-combustible surface.

Changes go beyond the buildings and land, such as purchasing insurance while preparing and testing an evacuation plan. How will we receive emergency alerts? Do we understand what they mean and how to respond? Do we know when to leave, how to travel, where to go, and what to take? How will we contact family members who are elsewhere when we evacuate? Have we considered taking irreplaceable and sentimental items, essential documents, and enough hygiene products, medications, and medical aids? Finally, is our address marked clearly on a non-combustible pole and reflective surface visible from both directions along the road through smoke so that emergency services can quickly locate our property?

These actions mean taking personal responsibility, with every family actively pursuing their plan, implementing it themselves for themselves. But nothing can happen in complete isolation, so another Wildfire Partners' principle is to work with neighbours, to compare notes, to exchange advice, and to collaborate on changes needed along property borders or roads. Wildfire Partners encourages participants to organize local meetings and to get their neighbours involved.

In the mountains, neighbours are not side by side and might not even be within shouting distance. They can be a ten-minute walk down the avenue or sited on the next ridge, distances which wildfires leapfrog in an instant. After the Cold Ridges Fire, some of the neighbours of those with houses following the Wildfire Partners programme returned post-evacuation to find ashes where their homes had stood. For months afterwards, with the scars of burned trees barely starting to be covered up by nature's renewal, hammering and sawing could be heard across the landscape as the neighbours rebuilt, living elsewhere until their new homes were complete.

Wildfire Partners participants never let down their guard. Fires can spark at any minute of any day, especially in the summer. Too often, there is barely enough time to leap into a car to escape.

During the Cold Springs Fire, one lucky resident dodged the flames on a horse, emerging uninjured. When evacuation means skedaddling immediately by any means possible, once we smell smoke, it is far too late to consider fire-resistance measures around our land and buildings.

Instead, overlooked by the skeletons of scorched trees guarding hillocks around their houses, home owners who do not want their possessions to ignite clear brush and debris, clean their gutters and eaves, trim the grass, thin limbs and branches from trees, maintain aspen which burns less than the lodgepole pine they remove, and rip out vegetation that encroaches close to the house. Some are self-employed, running businesses they founded, and the time they spend on avoiding wildfire destruction detracts from time spent with their clients. Nevertheless, in the end, the fire-related endeavours cost far less than losing everything in a few, sizzling minutes. Accepting the quality of life of living in the airy forests among the snow-capped peaks means the continual effort that comes of living in a burn area. Even so, as Wildfire Partners repeats: 'There are no guarantees'.

Wildfire hazards exist around the world. The triggers, intensities, and spreads can be forced as much by vegetation management, people management, and land use decisions as by the environment delivering lightning, wind, air temperature, and humidity. These points on reducing wildfire vulnerability neither condemn nor condone the choice to manage forests and fire. They emphasize that hazard modification techniques always yield advantages and disadvantages—as does avoiding changes in hazards. Addressing vulnerability, no matter what the wildfire hazard, must always be the focus of action to avert wildfire disasters.

Fire can be quenched by water. Water, too, has a role to play in nature and in destruction. Standing on Singapore's Marina Barrage, where five rivers meet in a bay that flows out to sea, gives a sense of the lengths (literally and figuratively) to which we are willing to go to try to control nature. At 350 metres long, it is just shy of the world's longest cruise liners, forming an imposing, concrete end to a stroll through the Gardens of the Bay park in the lustrous heat and humidity.

In desalinating Marina Bay behind it, the barrage provides a water supply for the city state. It stops many high tides from flooding low-lying areas of the city and drains excess rainwater from these same locations during deluges. The entire area has become a tourist and recreation attraction, for walking along coastal paths, boating in the bay, or relaxing in the shade of the adjoining garden's trees or the barrage's visitor centre.

The grassy rooftop above the visitor centre provides magnificent vistas of the barrage and Singapore's eclectic downtown architecture. The scale of engineering in Singapore city becomes conspicuous, from the coasts, up the rivers, and around the centre. This engineering and urban development placed people and buildings in the way of floods, and exacerbated those floods, so the city has now sought to alleviate these hazards through the barrage.

Around the world, river and coastal engineering dictates where the water goes, how fast it flows, and the power of the waves and currents. Human interventions to influence areas of flooding can be in the form of embankments or walls as well as dredging, building groynes, tailoring coastlines, and re-forming the bends of a river. Whether the water falls from the sky, melts from mountain peaks, or encroaches from the sea, we have spent millennia separating ourselves from it.

These endeavours make sense. Daily life would not be easy if we were continually flooded. Many peoples around the world, from Guyana to Myanmar, thrive in houses on stilts or on boats. In a different context, London and Cambridge in England have flourishing groups of boat dwellers, enjoying life on their canal or river. But not everyone desires this lifestyle. Plenty of infrastructure functions best when not immersed every so often. There is nothing inherently wrong with reshaping our environment to try to stay dry with the added advantage of channelling water for irrigation and drinking. The question is: how much does it really reduce flood risk over the long term?

Imagine that we live near a river which floods every few years. We get to know the water's cycles and we learn the signs of the river's highs and lows. We are cautious about storing valuables on our ground floor. We refurbish it so that it is easier to dry and clean after a flood, plus we make the electrics and plumbing water resistant. We even chat with our insurance company. We let them know where we live and ensure that we are covered for floods above the ground floor and for non-river floods, such as a pipe burst, a bathtub overflowing, or rainwater pelting through open or broken windows. As part of the deal, we agree not to claim for any ground floor inundation from the river. In short, we learn to live with the regular floods. We accept that we gain from living beside the river, with the cost of making some adjustments to our property and life alongside a bit of disruption. We are ready to deal with the typical river water ourselves while having backup for other types of flooding or unusual river extremes reaching above our ground floor.

Now imagine that we construct an embankment along the river, halting the regular flooding. We look forward to staying entirely dry while enjoying the river's amenities, the view, and the walking path atop the embankment. We need to repaint our house, but now we do not need to worry about water-resistant, easy-to-clean finishes, or about maintaining the water resistance of the electrics and plumbing. We start to use the ground floor exactly as the rest of the house, displaying artwork and filing the family's passports and wills in our ground floor study. We delay renewing our flood insurance, balking at yet one more bill to pay, eventually leaving it buried beneath a stack of paperwork.

One day, a bathroom pipe bursts when we are at work, pouring water into the ground floor for hours. Or a 'reduce taxes' government is elected, so their first budget cuts all monitoring and maintenance of the embankment. Perhaps, during a storm, a river boat collides with and breaches the embankment outside our home. An extreme storm could overtop or undermine the protective barrier, with a rush of water slamming into our property as the embankment crumbles.

In the final scenario, we would have been flooded even without the embankment. But if it had never been built, we would have been ready for the flood hazard and we would have reduced our vulnerability. The embankment's presence lulled us into losing our flood risk knowledge, permitting vulnerability reduction measures to lapse. We see the embankment, we are told that it separates us from the river, and we assume that we are protected from floods. This false sense of security increases flood vulnerability over the long term by eliminating some small-scale flood hazards in the short term. Without other actions to tackle flood vulnerability, we create a higher flood risk. The absence of an embankment would also curtail fast-flowing floods smashing into our walls. The water would typically rise slowly as the river swells and spreads out. The collapsing embankment could add a significantly dangerous component to any flood.

The important point here is to admit that we can do something about our vulnerability and stop disasters, no matter what the hazard or what we do to the hazard. But we must make the decision to do so. A mindset of prevention accepts the advantages and limitations of the river embankment. We still need a flood-resistant ground floor, flood insurance, and action plans in case of different flood types. We also need to understand how the embankment might have changed the flood regime, outdating our knowledge.

These points on reducing flood vulnerability neither condemn nor condone the choice to construct the embankment or to otherwise engineer the river. As in the case of wildfire, they emphasize that hazard modification techniques always yield advantages and disadvantages—as does avoiding changes in hazards. Addressing vulnerability, no matter what the flood hazard, must always be the focus of action to avert flood disasters.

Managing Ourselves

Canvey Island sits downstream from London, England, in the middle of the Thames Estuary leading out to the North Sea. Today, its population is nearly 40,000, and they are so proud of their island that an independence party has been born—demanding independence from its mainland borough council that is, rather than from the United Kingdom.

Canvey Island is artificial. Remnants of both Celtic and Roman habitation have been unearthed, but shifting land and sea over the centuries slowly reduced the island's habitability. During medieval times, it was marshland, frequently flooded with saltwater, and grazed by livestock. In the seventeenth century, Dutch engineers led by Cornelius Vermuyden drained the island, with the first sea walls built in 1623. The modern era's first Canvey communities came from Dutch settlers.

Rapid population expansion did not take place until the 1920s. A population of 1,795 and about 300 buildings are indicated in statistics from 1921, while by 1927, more than 6,000 people and nearly 2,000 buildings are listed.[4] This period marked Canvey Island becoming a retreat from London, particularly from the East End, for taking seaside holidays and spa breaks. Today, Canvey offers a getaway from London property prices as well as being a quiet retirement locale.

Building in a floodplain brings consequences. Draining water from soil compacts the land and the weight of buildings pushes it down further. Starting at sea level as a marsh before it was drained, much of Canvey's land soon descended below the mean high water mark of the River Thames. The night of 31st January/ 1st February 1953 brought the Thames and North Sea to Canvey. A tempest blew off the Atlantic, rounded Ireland, and headed across Scotland. The Irish Sea ferry *Princess Victoria* sank, killing at least 130 people. Around nineteen other deaths occurred on the waters around Scotland before the storm barrelled south across the North Sea.

A storm surge is coastal flooding that combines two phenomena. First, low atmospheric pressure in the centre of a storm pulls the sea surface upwards. Second, strong winds pile up sea water at the shore. The stronger the wind and the greater the distance over which it blows (the 'fetch'), the more water ends up at the coast, raising sea levels. Storm surges can inundate coastal properties with several metres of water.

Some tidal ranges exceed the storm surge height. If the storm surge arrives at low tide, it might look as if the tide fails to retreat. If the storm surge looms at high tide, then the tide appears exceptionally high. Tides display monthly and annual cycles. It was the misfortune of North Sea settlements that the 1953 storm surge swept along England's east coast when the tide there was near a maximum of the daily, monthly, and yearly cycles. Over 300 people died on land in England. Perhaps another 100 or more perished on boats across the North Sea while Belgium's official death toll reached up to twenty-two. The storm surge's full fury was saved for the Netherlands, where large areas lie below the high tide mark. There, 1,836 people succumbed to the cold and the water during that bitter night.

In England, the worst-hit area also rested below the high tide mark: fifty-nine are now said to have died on Canvey Island, although fifty-eight was the traditionally reported death toll for decades. The waves breached walls and engulfed streets up to bungalow rooftops. Just a bit higher and the water would have washed away dozens of people sheltering on top of their homes.

The UK government's response to this disaster involved a thorough evaluation and reworking of the strategy for stopping a North Sea storm surge from becoming a flood disaster.[5] Unwillingness to move away from floodplains, which would have meant abandoning most of Canvey, led to a focus on engineering coastlines. Separating land from water was accepted as protecting people and property in the Thames inundation zone, in which large expanses of London are sited. An eventual outcome of the government's review was the construction of the Thames Barrier, alongside raising and strengthening walls where the Thames runs through the city. The expectation that London is now immune from storm surge and Thames flood disasters is one factor driving expensive riverside developments. The high-rises of the financial centre of Canary Wharf ascend from the revealing place names of Mudchute and Marsh Wall.

Signs of the original ecosystem are not limited to place names. Across the river, on the Greenwich peninsula, an ecology park presents a wetland to educate locals and visitors about the nature which should be there. Wet areas which would previously have adjoined the river, soaking up rainwater and providing room for a storm surge, have now been drained for high-rise flats with prices exceeding £1 million. The Thames in London has nowhere to spread out except into the infrastructure.

The evidence of the changes made to the river remains. The Thames Path is a walking trail that goes right through London, allowing people to wander along the banks of the river cutting the megacity in two. Dodging the crowds leads us past landmarks such as the Houses of Parliament and the London Eye. Trashy novels and second-hand books can be picked up from the open air market at South Bank while Prime Meridian Walk marks its namesake of zero degrees latitude with mosaics in the pavement as it angles north from the river banks.

Walk past the current site (close to the original) of Shakespeare's Globe Theatre near the Tate Modern art gallery in Southwark and it becomes evident how much the river is controlled in central London. Sheer walls confine the flow, with the tide revealing and immersing scattered, pebble-strewn 'beaches' that can trap people as the water rises. The Royal National Lifeboat Institution's busiest station sits in the shadow of Waterloo Bridge with a rescue crew on-site and ready to go 24/7, because the cold, fast-flowing Thames gives people only minutes to live if they fall in and few ways to climb up the slippery walls.

The Thames Path in London reveals the waterway's history of continual control and narrowing as the city expanded. Just south of Whitehall Gardens, between the Embankment and Westminster stops on the London Underground, a weatherworn plaque describes 'Queen Mary's steps'. Excavations in 1939 uncovered steps designed by Christopher Wren in 1691 for Queen Mary II, who used them to descend to a river terrace. Today, the steps sit over fifty metres from the bank of the Thames. To get between the two, you need to traverse a wide pavement, the two-way bicycle superhighway, the two-way traffic of Victoria Embankment, and a grassy expanse. This much of the River Thames has been filled in since the end of the seventeenth century.

Limiting the width of the river means that, for the same amount of water, the depth and speed will increase. Evidence for this appears further downstream, underneath Southwark Bridge. Pedestrians rush through the tunnel on the south side, enjoying or seeking to avoid music from the buskers making good use of the acoustics. Engraved in the tunnel is a description of frost fairs, carnivals held on the ice when the Thames froze. The last one was in 1814, after which new bridge designs and sustained river engineering quickened the tidal flow, inhibiting freezing.

We humans have shaped and altered the river, which means that we have made London's floods by constricting the river's water between walls. In other words, human actions have made London vulnerable to floods. While a storm surge driving up the River Thames or rainwater coursing down it has its origin in nature, the flood which central London would go through and the damage it would wreak would be of human construction, by putting extensive, expensive property on land which would otherwise have taken these floodwaters.

Back on Canvey after the 1953 disaster, a concrete wall sprang up around most of the island. Rising more than three storeys above ground level, it is supposed to stop Canvey from being flooded by a storm surge beyond the level of that in 1953. Once any wall is built, the work does not stop, since walls must be maintained. Canvey's walls display signs of deterioration with deep cracks emerging. The seals between wall slabs are starting to disintegrate, with small plants growing out of some of them. Along Canvey Island's south shore, leaving access gates ajar allows people to enjoy a riverside walk at their leisure. Closing a gate means that someone must align metal bolts attached to the gate with holes along the wall. Some of these holes are clogged by debris and some rubber seals underneath the gates do not fully close the gap between the gate and the wall. Some of the locks appear to be rusted.

Encircling the island with walls is an attempt to manage nature by keeping the water out. For low-level storm surges or storm surges occurring with low tides, the walls largely succeed: Canvey has not had a big flood since 1953. But no wall can be 100 per cent safe. If one of the walls collapses, breaches, is undermined, or is overtopped, then the flooding could be devastating. Few will be prepared for it, because they believe in the safety of the walls, as well as the right to occupy land that was previously part of the sea.

So how much should we try to manage nature's waters? The baseline is that no one wishes to see a disaster. Letting a North Sea storm surge trap people in their homes must be avoided, and that can be done by balancing measures to keep people safe. Walls are currently the main part of this plan: manage nature by controlling where the water goes and by separating the people from the water. People are kept entirely dry—until a storm surge arrives that is larger than expected, or the supposedly protective measures founder.

Another option, especially given the expectations of higher sea levels under climate change, is to return parts of Canvey Island to the sea. Marshland breaks the force of waves and gives room for the water to spread out. It also reduces the space in which people could live without guaranteeing that the inhabited areas will always remain dry. Another possible though large undertaking would be to raise the land, so that all infrastructure sits above the expected storm surge level. This is expensive, disruptive, changes the character of a community, and also has no guarantee of working. Wave, tide, and current action could gradually erode the raised land, although maintenance can assist as long as money is available. And a decision is needed for how high to go.

Leaving for the mainland in the wake of a storm surge warning is another option. Until the first bridge between Canvey Island and the mainland opened on 21st May 1931, ferries were the main route on and off the island. Two public roads now form the main connections with the mainland. They intersect at a roundabout, making it effectively one-route unless traffic flow measures applied during an evacuation force the roads to operate as a one-way system. Narrow lanes across non-public land potentially provide a third way off the island to the west, if the gates en route are left open and if a large number of vehicles would not damage the passages. But evacuation means leaving behind house and home. As with wildfires, evacuees must take with them irreplaceable items and be ready to rebuild immediately afterwards.

In the absence of a major storm surge disaster, retaining a high level of readiness is not easy. It means managing ourselves much more than nature. Thus far for Canvey and London, the decision has been almost exclusively to manage nature by separating water and land, rather than also managing ourselves. Others, though, have taken a different approach.

One homeowner along England's Essex coast has long known about the chances of floods and the damage which water does to an unprepared building. Rather than trying to reconstruct nature to avoid flooding, he renovated his house to make it easy to deal with water. He raised all the electric wires and sockets on the ground floor and installed drains. The water can flow in up to a certain height without knocking out the electricity, and then can easily flow out. To facilitate cleaning and reduce damage to his contents, the owner removed all carpets from the ground floor and coated the walls and floors with a water-resistant plaster used in swimming pools.

Would this work for everyone? Some people like carpets or prefer to have electric sockets near the floor. Finishes such as paints and plasters need to be selected carefully to avoid health hazards from off-gassing. Many would consider floor drains to be aesthetically displeasing. In the end, it is about managing ourselves in terms of knowing the options available and accepting the consequences of our decisions. A property owner could accept that, every time it floods, all carpets and electrics will need to be replaced, with alternative accommodation found for the months needed for cleaning and drying. Or only some flood resistance measures could be taken in order to balance flood-related disruption and recovery. Options exist, each with advantages and disadvantages.

Much of the hazard, then, is shaped by humans—as much as, or more than, by nature. This does not stop us cursing nature's malevolence when a flood inundates shops or a wildfire razes a school. The environmental events and processes originate in nature and so we call them 'natural hazards'.

The balance and integration of the processes of managing nature and managing ourselves to deal with hazards depends on values and preferences. In both London and Singapore, a particular approach was taken regarding floods, and changing it would be expensive and time-consuming. It might not even be politically and culturally acceptable. It is clear that, as with the early European efforts to deal with Australian bushfires, a focus on managing nature deals with the hazard without doing much to understand and tackle vulnerability.

Characterizing nature as dangerous and malfeasant, and hence needing to be tamed, typically creates an emphasis on hazards. Disaster vulnerability and risk then tend to increase. Hazards, though, are not inherent to nature, being regular and typical environmental processes and events; they become hazardous only when faced by an unprepared society. The potential damage which these 'hazards' can do is partly created by human design and management of the environment and our society. We permit, actively and passively, much of this damage to occur.

This manufactured hazardousness of nature masks many of the resources and opportunities brought by 'natural hazards'. Many people settle in floodplains and around volcanoes because the floods and volcanic ash enrich the soil with nutrients, yielding productive farming. Faults associated with earthquakes permit deep groundwater to percolate up to the surface, providing a lifeline in arid regions.[6] Many desert cities developed over faults, in locations which now seem far too hazardous for the infrastructure which we have chosen to construct there.

Nature has resources which we can use and manage, but inadequate or inappropriate approaches can make nature hazardous. Nature doesn't mind either way, as it is neither good nor evil. It is up to us to manage nature and, far more importantly, to manage ourselves to avoid exacerbating or creating hazards and hazardousness. We can live with and use nature's events and processes as resources without disasters happening, although it requires planning and preparation. To do so, we must admit and tackle vulnerability.

CHAPTER 3

THE STORY OF VULNERABILITY

We know how to build infrastructure to withstand most earthquakes and tornadoes. We know when tsunamis might come, since a hazard—perhaps a landslide, earthquake, volcanic activity, or meteorite strike—always precedes one. We have structures, materials, paints, and electrical equipment which can survive floods and sandstorms. Yet, despite this knowledge, we do not always implement what we know, meaning that such events result in injury, death, and damage to homes and infrastructure. Why do we not continually use the knowledge we have to avert disasters?

The explanations are many, varied, and complex. They reveal sordid tales of vulnerability: what it is, where it comes from, why it exists, what we ought to do about it, and why we do not always do what we ought to. They concern actions, behaviours, values, decisions, and choices—not just our own, but also of those with the power and resources to decide for others, with or without their awareness and consent.[1] The decisions occur continually over the long term, determining how society treats different groups, and how it governs, distributes wealth, and makes and implements choices.[2]

These wide-ranging actions and values create what is referred to in disaster studies as 'vulnerability'. The vulnerability that builds up results in known and usual phenomena of nature becoming hazardous to humanity, so that disasters occur.

Talking and writing about disaster vulnerability is easy; understanding it is hard. Who creates vulnerability? Who wields the power to stop vulnerability to disasters? Who should take responsibility for tackling vulnerability? The answer to all three questions is: everyone. From individuals through to intergovernmental organizations and multinational corporations, we are all responsible for reducing vulnerability and dealing with disaster risk.

Sorting through the myriad influences, people, groups, and organizations is complex. We face an exceptional struggle in trying to be comprehensive about vulnerability by connecting everyone and everything. But this quagmire must be faced, since we are all responsible for tackling the causes of vulnerability, though some people and groups have more choices, more resources, and more power than others. Documenting the creation and perpetuation of vulnerability is not simple either, and this chapter cannot meet the challenge. Still, it touches upon many of the key aspects of what causes vulnerability, where it comes from, and why it is created and continued. It aims to present the essential pieces of the disaster jigsaw, which together form a considerable part of the entire, grim picture.

Exceptions are prevalent. No group ever comprises entirely homogeneous individuals, so to talk of 'men' or 'people with disabilities' as coherent collectives conceals the extensive differences among the individuals and subgroups within each. Giving an overview of vulnerability for anyone, anything, or any collective leaves gaps which individuals and groups fill in according to their own biases and interests. Clashes arise as people close these gaps differently. Still, together, and because of rather than despite disagreements, we can depict as complete a picture as possible of vulnerability to disasters. Then, we can understand where vulnerability comes from, who is affected, to what level, why all this happens, and how it might be addressed.

Vulnerability by Numbers?

The scale of a disaster is defined by its impacts. The greater the number of people who are affected significantly, the larger a disaster is assumed to be. What is the significance of population numbers for vulnerability?

The slopes of Popocatépetl volcano, overshadowing Mexico City, are packed with expanding shanty towns. Some people move there because they believe that the city offers opportunities, often to become disillusioned. Others crave an escape from the violence, gangs, drugs, and poverty in their home towns, only to find much the same in the big city. Either way, they are poor and marginalized, with little choice but to move up the mountain's slopes into increasingly hazardous areas, boosting the population numbers around the capital city.

Meanwhile, in South Carolina, multimillion-dollar McMansions grace the streets of Hilton Head, with beautiful beaches a short wander along the same paths the ocean can take during a hurricane to wreck these same mansions. Hilton Head has had its recent share of storms. As hurricane Matthew swept by on 8th October 2016, fresh from its destruction in Haiti, docks were ripped out and sailboats were thrown ashore, combining with structural impacts to add up to over US$50 million of damage. Residents were evacuated for nearly a week. September 2017 saw a three-day evacuation and over 200 properties damaged by tropical storm Irma. A year later, Category 4 hurricane Florence was making a beeline for Hilton Head, ending up not curving as far south as expected, and then making landfall as a tropical storm. Little damage was reported around Hilton Head, although a mandatory evacuation order which was lifted after twenty-four hours led to confusion and interruption of business.

Hilton Head and Mexico City appear to form a striking contrast of affluence and poverty, of sailboats compared to gangs. The rich population of South Carolina can purchase insurance, can self-evacuate in private vehicles, can afford to take several days off work (and might be paid anyway if they take holiday or if their employer covers evacuation days), and can pay for alternative accommodation. Yet it is actually the same for the rich people of Mexico City: before and when a hazard manifests, they can afford to take care of themselves. Meanwhile, the poor of South Carolina face similar difficulties to the poor of Mexico City. They lack resources, job security, access to healthcare, and opportunities for improving their situation. All these represent disaster vulnerability, so disaster vulnerability is not just about the total number of people.

Losing a recreational sailboat impacts lives differently to having your only home ruined, whether you are Mexican or South Carolinian. Given that different people are affected dissimilarly by the same hazard, the proportion of people affected in various ways indicates diverse dimensions of a disaster's severity. Some days away from home followed by several weeks of clean-up might be the consequences for the richer people while the poor people stand to lose everything they have—including their lives. The overall severity of a disaster includes the proportion of people going through particular impacts in addition to the numbers affected.

Islands exemplify this point. Prior to its volcano starting to rumble in 1995—with volcanic activity continuing for two decades after—the Caribbean's emerald isle of Montserrat had a total population of around 12,000. Up to twice this number were killed during the earthquake of 26th January 2001 in Gujarat, India. If we consider only the total numbers, the disaster in India far surpasses anything which could ever afflict Montserrat. Yet India can draw on the resources of over a billion compatriots to assist. Montserrat could, and did, draw on its diaspora, still numbering in the thousands, keeping in mind that India's diaspora is orders of magnitude larger. Whether or not vast resources are allocated fairly (and they aren't) and whether or not vast resources are used for suitable disaster assistance (which they weren't) is another question. In Montserrat, these questions could not even be asked.

Since 1995, every resident of Montserrat, 100 per cent of the population, has been directly affected by the volcano and close to 100 per cent of Montserrat's pre-eruption infrastructure has been severely damaged or destroyed. On 25th June 1997, hot ash and gas clouds called pyroclastic flows barrelled down the volcano's slopes, killing at least nineteen people who were the main immediate fatalities from the volcano during twenty years of eruptions. This number is tiny compared to the fatalities in the Gujarat earthquake, yet as a proportion of the population affected, it would be equivalent to more than one million people dying in India.

Disasters in India are devastating in their own right and should never be downplayed, especially considering what each affected individual goes through. But the forms of vulnerability and disaster experienced in India and in Montserrat display major differences when the total number of people affected is compared with the proportion of the population affected. India's disasters far exceed anything in Montserrat in terms of total numbers of people affected, while those of Montserrat far exceed anything in India in terms of proportion of population involved. And the comparisons by number do not stop here. The proportions of people within different populations—grouped by age, gender, sex, sexuality, race, ethnicity, first language, or disability—should also be compared with respect to how they are impacted by a disaster.

Another dynamic affecting vulnerability as measured by numbers is the global trend of increasing life expectancy. People are living longer on average. What do age and ageing mean for vulnerability to disasters?

Children are generally less able to take care of themselves than most adults. Children lack the experience that comes with years of learning, watching the world, making mistakes, and surviving those mistakes. As children mature and learn, their physical abilities peak as young adults. Physical ability is a small component in surviving a disaster; it helps, but should never be relied on. Without previous thought on how to react, the person who can run faster or climb a tree might make it off the beach as the tsunami strikes. But not always, since the water can smash through buildings faster than people can run and a skilled climber could die if the tsunami overtops or knocks over the tree. A more physically fit person might survive traumatic injuries inside a building which collapses in a tornado or earthquake. If they are struck on the head or cannot stop a spurting artery, then only urgent medical care might save them. In a flood, strong and weak swimmers can easily succumb to hypothermia, be knocked unconscious after colliding with obstacles or debris in the water, be sucked under by the flow, fatigue themselves by swimming, or be contaminated by oil or chemicals in the water. For age and physical ability, as many exceptions as rules permeate disaster survival.

Many children have defied the odds against survival for days before being pulled from the debris after an earthquake or mud-flow. On 17th August 1999, a powerful and shallow earthquake struck parts of Turkey, killing around 17,000 people. Into the fifth day of rescues, four children aged 9 to 11 were extricated from collapsed buildings, one after a seventeen-hour operation to free him. On the same day, a 95-year-old woman was pulled from the rubble, so survival is down to circumstances too. The following day, the sixth after the earthquake, a 4-year-old boy was rescued, suffering mainly hunger and thirst. His mother, having been pulled out earlier, survived, although his father and three sisters had already been killed by the collapsing building. When an earthquake rattled southern Italy on 31st October 2002, an entire primary school class was wiped out. The roof of their relatively modern school collapsed in San Giuliano di Puglia, killing twenty-seven children and their 47-year-old teacher. Neither youth nor middle age saved them from a badly constructed building. The only two other deaths from this disaster were people aged 54 and 91.

Epidemics display different age distributions for the people most affected. Young people are much more at risk from meningitis while older people are felled more by influenza. We gain immunity to many diseases after a single infection, so if we survive it, then we are at much lower risk in future outbreaks, no matter what our age. We can't let down our guard though, since we might still be vulnerable to different strains of the same disease.

Those of us who successfully navigate the dangers of our early years gradually lose our physical and mental faculties as we get older. Many earthquake deaths are from heart attacks, when people are terrified by the shaking or stressed in the days afterwards. On 7th June 1931, the strongest measured earthquake to rattle the UK so far struck at moment magnitude 6.1 underneath the North Sea off the coast of England. The British Geological Survey reported one death in the UK, from a heart attack in Hull. Cardiac maladies affect the old more than the young, as do other ailments such as osteoporosis. Weakened bones mean an increased chance of injury.

The human body's ability to regulate its temperature also diminishes with age, so we become more vulnerable to hot and cold extremes as we become older. The 1995 heatwave in Chicago and the 2003 heatwave across Europe are infamous for the high proportion of elderly people who died. These deaths were not purely from the temperature-related physical attributes of age. Many of the elderly were socially isolated: no one knew they were in distress, they could not afford fans or air conditioning, and they suffered from the failure of public services to respond to an unfolding crisis. In Chicago, legitimate fear of crime precluded many from responding to door-knocking volunteers and municipal employees who were trying to check on their health. The thought of being robbed deterred them from leaving their homes to rest in cooler common areas of apartment blocks or in shaded areas outside. This complex array of issues affected elderly people far more than others. Vulnerability to the heatwave disaster was not fundamentally about physical temperature regulation in the human body. It was about wider social processes exposing a certain proportion of the population to protracted, lethal heat and humidity. If we solve the social problems affecting many of the elderly, then we could be less concerned with the physical problems of age. This is vulnerability by numbers based on age.

Conversely, when it comes to the young, could teaching overcome some age-related aspects of vulnerability by conferring knowledge and experience? Children develop language and communication skills through imitation and education. Continual interaction with people is one major component, as is nutrition and environment, since children need to eat appropriately and also avoid pollution for their intellectual and cognitive functions to develop fully. Key messages about society are communicated in various ways as a child matures, engraining a sense of culture, values, and norms. This can involve ways of coping with hazards. Carefully explaining to a toddler that cars can easily be carried away by floodwater, and that washed out roads can be hidden by floodwater, might not instil in them that we should never drive through floodwater. Rational argument might not even convince teenagers to overcome their belief in their invincibility. But seeing adults around them refusing to take their vehicle into floodwater, in addition to discussing and explaining the reasons behind this behaviour, can have a powerful influence on the child and their future actions. It might, naturally, also induce rebellious teenagers to try driving through floodwater. More than one medium of communication must be used to reach different audiences with reinforcing messages.

Not all audiences use the same media for communication. Some people are born with different forms of visual and hearing abilities, while for many others visual and hearing disabilities appear and worsen with age. Understanding the numbers of people with various capabilities contributes to understanding vulnerability by numbers and how to overcome it.

Hearing disabilities should be straightforward to deal with, since audio channels are not the only mechanism for relaying warnings. Text messages, video messages, and visual news flashes for websites and television contribute to ensuring that lack of hearing does not mean lack of information. Hearing aids, from cochlear implants to simple amplifiers, suit some people as do vibrating pagers to wake them up with alerts. Disaster vulnerability in this context is not about lack of hearing, but about lack of access to alternatives. Visual disabilities should be similarly straightforward to overcome by using audio sources, providing glasses, making large print available, or operating to remove cataracts.

Extensive, creative approaches exist to assist people in understanding and dealing with disaster-related topics, no matter what their age or their health-related concerns, whether linked to ageing or not. In any case, age brings knowledge, wisdom, and experience which should be advantageous, although this is not always evident. Having witnessed a lifetime of disasters and being aware of the dangers could motivate people to prepare, although it can backfire through overconfidence based on presumed understandings. When warnings were issued for New Orleans in 2005 as hurricane Katrina bore down on the city, some elders narrated how they survived hurricane Camille in 1969 and hurricane Betsy in 1965. They had been through it all, so why should they worry about Katrina? This attitude is not necessarily wrong, since the hurricane itself is only the hazard, not the true danger of a disaster seen via vulnerability. If we survived one Category 4 or 5 storm, surely we should be able to survive another? Except that, in the meantime, we have changed the environment and society, altering vulnerability.

The four decades from 1965 to 2005 had witnessed the destruction of many of Louisiana's wetlands. More power from a hurricane's storm surge could now batter the inhabited coastline. Levees had been shored up around New Orleans in the decades prior to Katrina, but they had never been tested in severe hurricane-related flooding. Those levees failed during hurricane Katrina, inundating parts of the city. In addition, urban development produced faster and increased runoff into water channels and low-lying areas during rainfall as the city's population and population density had increased.

Hurricane Katrina as a storm might or might not have been different to previous storms experienced across New Orleans. Hurricane Katrina as a disaster was vastly different from Betsy and Camille, and the difference was due to increased vulnerability. The wisdom of the elders can fail if it doesn't account for all knowledge or for changes from the past. In such cases, a disaster can affect far more people than mistakes made by the young, because those with power and in leadership positions tend to be older, though, as always, there are exceptions.

The advent of smartphones and social media has thrust many young adults into the world spotlight. Mark Zuckerberg, born in 1984 and founder of Facebook, and Laura Bates, born in 1986 and founder of Everyday Sexism, have accumulated influence, attention, and power to the extent that they are world leaders. When she was 17, Malala Yousafzai won the Nobel Peace Prize for promoting children's right to education. Democratic elections around the world have given leadership roles to politicians across a spectrum of ages in recent years. Donald J. Trump was elected president of the USA in 2016 at the age of 70, a youthful contrast to 92-year-old Mahathir Mohamad who was elected two years later as prime minister of Malaysia. In between, in 2017, Jacinda Ardern won New Zealand's election to become prime minister at the age of 37, still six years older than Sebastian Kurz, who was elected Austria's chancellor in the same year.

Despite the mantra that vulnerability by numbers means higher vulnerability with age, for the many world leaders who are elderly, their power provides them with options to improve conditions for themselves and society, if they choose to do so. They do not always choose to do so. Many leaders, irrespective of age, make choices which actively increase disaster vulnerability for themselves or for others. These choices are political, indicating how physical differences such as those arising through ageing might be much less influential on disaster vulnerability than political choices. The physical and mental changes, positive and negative, which come with age are never denied, but they cannot overcome the dominance of political choices in influencing all vulnerabilities. In the end, vulnerability by numbers in terms of how old we are or our abilities, does not sufficiently account for vulnerability. If we solve the social problems of vulnerability, then we could be less concerned with attempts to ascribe vulnerability to age.

Given how age does and does not influence vulnerability, we can again see how it is not just the total size of a population, or indeed even a population's age distribution, that affects disaster vulnerability by numbers. Each is a factor, yet does not stand alone—and the same applies to population density.

We often hear that moving to cities creates vulnerability and that urbanization, particularly along coasts, is a major cause of disasters. The reality, as with all the other factors of vulnerability by numbers, is more complex. Urbanization implies increasing population numbers concentrated in a comparatively small area, so more people are affected by a hazard in that area. Epidemics are an example of high population numbers and densities potentially promoting the spread of a hazard, that of microbial pathogens. But high population numbers and densities have another side; it means more people are available nearby to assist quickly in a disaster. In many countries, cities have the best equipped, most experienced, and fastest-responding emergency services, as well as logistics officers for preparing and then distributing supplies, and good medical care. These all help to prevent and deal with disasters—unless they are put out of action in the disaster, a situation which can be avoided through preparedness and practice. If natural hazards occur regularly in a place, then we will be more used to them and hopefully more inclined to prepare for them and know how to deal with them, irrespective of population numbers, population densities, urbanization, or the nature of the hazard.

Singas Village in Papua New Guinea sits by the Markham River. The 300-or-so residents know that the river floods every rainy season. They do not wish to move, and why would they, since the river provides fish, clay for pots, and water for drinking and irrigation. Instead, they have changed their lifestyle to deal with floods, building mounds on which to site houses and then constructing their houses on stilts, to stay dry above the floodwater. Their agricultural plots include crops which can survive flooding alongside drainage to keep the worst floodwaters away from their food. They store food and collect drinking water to avoid hunger and thirst during floods.

With a regularly occurring hazard, we can choose to use our experience to reduce our vulnerability. Less common hazards could mean that we do not look beyond the familiar, at what we could and should do to help ourselves in the case of rare hazards, so our vulnerability increases. Irrespective, if vulnerability is addressed, then the rarity (or presence/absence) of a hazard can be immaterial. This is the case for coastal cities. Coasts do not necessarily have a larger variety of hazards than inland locales, because tsunamis, waves, storms, and changing water levels affect freshwater areas too, such as a rockslide into a lake or a river swollen with runoff. If moving to a coast takes us away from mountain hazards, such as volcanic eruptions and rockfalls, then the balance of hazards might lessen. The cities of Castries, St Lucia and Hong Kong are both coastal and mountainous, bringing an impressive variety of hazards. Nevertheless, if vulnerability is addressed, then a hazard is far less likely to lead to a disaster.

More than the parameters set by nature, the design of a city can affect both hazards and vulnerabilities. The layout can inhibit or support evacuation, given that areas of high population density could be harder to evacuate if large population numbers need to negotiate tight escape routes, although affluence plays a big role too. Those without private means of evacuation might be more inclined to seek and board organized mass transport, which can quickly move a lot of people out of the hazard zone. People who can afford private vehicles might depart in them but too many cars on the road quickly creates tailbacks and increases the chances of crashes.

In the USA, motorways are made into single-direction roads to support evacuation as hurricanes bear down. During hurricane Rita in 2005, this was initially not done in Texas, so people evacuating from the coast were caught in hours-long tailbacks. Cars broke down and ran out of petrol, many died in the heat, and a bus of medical evacuees burst into flames, killing around twenty-four in the vehicle. The number of people living in the area, coupled with poor evacuation planning, combined in a deadly fashion.

The expansion of cities upwards through high-rise structures changes hazards—for instance, by affecting wind patterns around the buildings and the earthquake shaking experienced at the top—and also vulnerabilities. Evacuating tall buildings cannot happen rapidly, but upper floors can be refuges in floods or slides, provided that the buildings do not collapse. Being inside structures which are properly designed, built, and maintained, no matter what their height, can be the safest places for surviving a hazard.

Cities change hazards in other ways too, with urban heat islands and wind tunnels affecting dwellers on a daily basis and exacerbating (though not creating) extremes. Roads, pavements, and car parks can hasten runoff into rivers, inducing floods in locations expected to remain dry. Conversely, green areas ease flooding and are often a feature of cities, cultivated to increase quality of life. All areas—urban, rural, and those in between—have vulnerabilities and abilities to tackle these vulnerabilities. A choice can be made to pursue paths towards reducing vulnerability which are not based on urbanization, population densities, or population numbers.

Vulnerability by numbers does exist, yet we can make choices to create, to control, or to reduce vulnerability irrespective of the numbers. The story of vulnerability includes vulnerability by numbers and much more.

Vulnerability by Ideology?

The region along the US Atlantic coast, around the Caribbean, and throughout the Gulf of Mexico, is designated the 'hurricane belt', for the obvious reason that hurricanes wander within this zone. For an equally obvious reason, June to November in these areas is called the 'hurricane season'. Hurricane names form an alphabetic sequence in English, starting with A, so the 2020 storms are listed as Arthur, Bertha, Cristobal, Dolly, and so on, skipping less common letters in English such as Q and X.

Coalescing over the open ocean, the winds of the most powerful hurricanes blow at 300 kilometres per hour, equivalent to the top speed of Eurostar trains. Gusts within a hurricane are much faster. The most damaging aspect of hurricanes is typically not wind, but water. Storm surges, just like in the North Sea, sweep over low-lying barrier islands and clamber up beaches. A storm surge in 1900 put Galveston, Texas, underwater, killing over 6,000 people and becoming the deadliest land-based hurricane in US history to date. Today, as hurricanes make landfall, dramatic footage of waves crashing ashore and washing away roads do not always deter foolish surfers and swimmers, many of whom never return from the swirling sea.

Nowadays, often more lethal than storm surges is the rain. Hurricane Harvey in 2017 deposited the most rain from a single storm yet recorded in the continental US. Some places received their average yearly rainfall during Harvey. Given that the hurricane season is the rainy season, any storm can lead to damage and interruptions if we are not ready for it. A storm with winds just below hurricane force is termed a tropical storm, and it too brings a powerful combination of wind, storm surge, and rain which can be just as deadly.

In 2005, while the world remained captivated by the aftermath of hurricane Katrina, which flooded portions of New Orleans, tropical storm Stan (later a hurricane for about nine hours) unloaded torrents across Central America. Katrina and Stan each yielded about the same number of immediate deaths, between 1,500 and 2,000, even though the stories and images from Katrina generated far more attention than those from Stan.

We need to be alert outside of the official hurricane season as well. Following tropical storm Stan, the 2005 hurricane season did not end until January 2006. So many storms formed during 2005 that no common letters in the English alphabet remained to name them, so the Greek alphabet was used. The season's finale, tropical storm Zeta from the Greek alphabet's sixth letter, appeared in the middle of the Atlantic on 30th December and lasted until 6th January, fully dispersing on 7th January. In 2017, far from land too, Arlene remained a tropical storm for around twenty-four hours on 20th to 21st April, more than a month before hurricane season officially began.

The above examples demonstrate that we need to be prepared to deal with the variability of nature. We know that hurricanes arise during hurricane season and sometimes beyond. We must be ready for more than half the year, which leaves the remaining months to prepare. In other words, hurricane readiness never starts and never stops; it has to be an ongoing process, a part of everyday life. Aiming to avert disaster is a continual, year-round task. And it works. By all reckoning, Harvey's immediate death toll was under eighty-five, a tiny fraction of the lethal devastation wrought in 1900 in the same area. The same is true of tropical storm Allison in 2001, which flooded Houston as a bungalow-high storm surge washed over Galveston, yet immediate deaths were two dozen lower than Harvey's.

In these storms, thousands of lives were saved through preparedness, warning, evacuation, rescues, and response measures. We can take action to survive nature's extremes if we choose to do so. The decisions not to do so typically take place over the long term, with machinations wider than most individuals can fathom or address. They involve broad and deep political choices, leaving little blame to be placed on what individuals do or do not do as a storm approaches their abode—especially given that it is far too late to address the baseline issues once a storm system starts rotating.

Tackling these baseline issues still has a long way to go in Texas. Allison and Harvey still killed, devastated Houston and surrounding areas, and turned the lives of millions upside down through floods, interrupted livelihoods, and disruptions to children's schooling. Since statehood, Texas has often in effect encouraged flood vulnerability, being dissuaded only briefly and infrequently by disasters over the decades. For Houston, so much could have been achieved before either Allison or Harvey had the city in its sights, especially considering the experiences of past storms.

An elementary rule is that not building on floodplains helps us to stay dry. Yet all locations experience some hazards, meaning that avoiding all possible threats of damage is impossible. Instead, a balance of potential hazards can be sought, while designing and building for this balance of hazards, which means tackling vulnerabilities for whatever combination of hazards might be expected and beyond. Techniques include land use planning and zoning regulations, which for Houston have long been slack.[3]

Yet we know how much we could do to develop and live in flood-prone locations without creating flood disasters. Car parks, streets, and pavements can be permeable, allowing rainwater to soak through them rather than running off immediately to pond in the low points. The city could have supported cheap, reliable, frequent, and safe public transport, to reduce the need for multi-lane roads and huge car parks, which blanket green spaces. Buildings and roads can be orientated to funnel rainwater into low-lying areas designated to catch and store it. Examples are parkettes that could turn into temporary reservoirs, ravines to be used for walking and cycling instead of living, and lakes or ponds with room to expand which could otherwise be places for picnicking and boating.

Some tree species are good at soaking up water, although one challenge is then that during rainfall shortages, they might suck up too much, exacerbating a drought, or else die. In any case, planting non-native species is not typically a good idea because they might significantly harm local species and local ecosystems. Meanwhile, any storage areas and greenways have a limit to how much water they can redirect or retain, so a possibility inevitably exists of this limit being exceeded.

At the time of hurricane Harvey, Houston was implementing many initiatives to support green space, to improve runoff management, and to reduce ponding of surface water. The city was also fighting to overcome an antithetical legacy generating vulnerability. The city's population had increased by 40 per cent since 1990 while the state, like many other places in the USA and around the world, displayed ingrained racism and desperate social inequities that make people vulnerable.

Voting records across Texas often favour lower taxes and oppose tackling prevalent social and economic disparities. Opportunities for marginalized people to assist themselves can be reduced. These choices create and maintain the vulnerabilities which caused the storms to become disasters. Voting for creating disaster vulnerability is an ideological choice and voters have the right to make these choices. The implications are foreseeable and are known, becoming manifest when Harvey swept through in 2017 and matching the long-standing reality of disasters across the state.

No hurricane tells people what taxes can and cannot achieve. No wind speed confers racism. No flood depth selects an oil-based boom-and-bust economy. No storm surge dictates where to settle and how to build. Changing storm regimes, due to climate change and natural variations, do not force people to vote in a certain way. All these creators and reducers of vulnerability occur through human choices; they represent vulnerability by ideology, understood here in the broad sense, as referring to all systems of ideas and doctrines ingrained within society and driving prevalent values.

Texas in 1900 and before, as well as in 2017 and after, chose to create its own vulnerability to hurricanes. Those with power, resources, and privilege have always had options to reduce this vulnerability and to avert the disasters. None of the storms striking Texas in 1900, 2001, and 2017 were disasters in themselves. The disasters arose from the deaths, destruction, and interruption of lives and livelihoods. Where action was taken, the disaster was far from as bad as where little was done in the hours and decades before each storm.

This insight, of vulnerability created more by society through ideology than by physical factors, applies also to gender—or, using the correct term, sex. 'Gender' and 'sex' are not interchangeable; they represent different attributes, and the dominant binary division of males and females is far from the reality of how many view and live their gender. Dividing men/boys and women/girls is typical of how disaster data have so far been reported. Research is ongoing to discern better the experiences across different genders of dealing with disasters.[4] At the moment, the majority of data and analyses which address gender or sex focus on categorizing people as being either male or female based on physical attributes alone— and the differences in physical attributes should never be denied.

After the 26th December 2004 tsunamis pummelled coastlines around the Indian Ocean killing over 200,000 people, Oxfam compared male and female deaths.[5] They tallied data from villages in Indonesia, India, and Sri Lanka, finding that fatalities among women consistently outnumbered those among men. In some villages, females comprised 80 per cent of those killed.

The conclusion is not that women are more vulnerable to tsunami disasters than men. Rather, these statistics indicate that society creates roles based on gender which lead to different tsunami vulnerabilities for men and women. In some villages in Aceh, the tsunami struck as women were at the shoreline, waiting for their fisher husbands to return with the morning's catch. Had the tsunami hit earlier, the men would have been killed as they launched their fishing boats. Had it hit later, the men and women would have been landing the catch, leading to more equal death rates. Meanwhile, in one Sri Lankan area, women were bathing in the sea as the waves rolled in, leading to higher death rates among women than men, which would have been different if the tsunami had hit a few hours earlier or later, when the women were not in the water.

The largely artificial and cultural separation of men and women in society, and the largely artificial and cultural creation of gender-based roles, leads to vulnerability based on gender which, in turn, can result in different death rates for males and females in disasters. The impact of gender-related differences can cut deeper than day-to-day roles.

In many tsunami-affected locales such as southern India, men are more commonly taught to swim than women. Despite swimming ability not always helping to survive immersion in a tsunami or a flood, it does lend a familiarity to being in water and an understanding of how to attempt to survive. In some places, women might not be permitted outside the home without being accompanied by a male relative. Even if a warning had been issued and the evacuation routes were known, without a male relative around women might have chosen not to evacuate.

The type of clothing which women are expected to wear in many of the tsunami-struck villages is also not helpful for moving quickly on land or in the water. Because of social and cultural norms, the women would never consider wearing different or fewer clothes in order to escape with their lives. Dying is preferable to being seen in public without what is considered to be proper attire.

Many households around the coasts of countries such as India and Bangladesh never have enough food. The household's woman (or women) cooks and places a meal on the table. The man of the household enters, eats what he wants, and leaves, after which the woman (or women) gives up a portion of her share to ensure that the children eat enough. Women then become hungrier, more deficient in nutrients, and physically weaker than men. They are less capable of reacting during crisis situations, not because of factors inherent to being female, but because of the ways in which they are expected to live as females.

These descriptions apply beyond the 2004 tsunamis, being seen in many other water-related disasters around the Indian Ocean. In April 1991, a cyclone made landfall in the Chittagong district of Bangladesh. Estimates of the death toll range from more than 67,000 to double that figure. Females died far more frequently than males across all age groups. While the difference was not huge for teenagers, four to five times as many women died than men among young adults and the middle-aged. The social status of women and codes of behaviour expected of them by virtue of their gender—that is, social norms and expectations founded in ideology and culture—influenced their vulnerability more than biological differences.

Many other factors are likely make women more vulnerable to disasters than men around the world. These factors are not well evidenced for disasters, even though they are known for other areas of life. Women might fear sexual violence along evacuation routes or in disaster shelters. Social rules in many places preclude women, especially if unmarried, from sleeping in close proximity to men—an all-too-common situation when there is a lack of pre-paredness for disaster evacuation and sheltering. Some societies still refuse to blame a perpetrator for sexual assault. Many cultures ostracize women who are menstruating, making it difficult for them to seek and remain in safe places when they have their period. The objectification of women and their bodies, fear of women's anatomy and bodily processes, and the normalization of psychological, physical, and sexual violence against women are forms of rampant and ingrained sexism and misogyny which represent a major factor in substantially increasing females' vulnerability in disasters.

Little data exist on these factors in relation to disasters because they have not been fully accepted as relevant or as important research topics. Pregnancy is also debated. A pregnant woman undergoes physiological changes which increasingly inhibit mobility under her own power, possibly limiting some of her physical responses to hazards. Debate continues over whether this situation creates unavoidable and inherent vulnerability in pregnant women (based on physical characteristics) or whether society's lack of caring and support for pregnant women is the true cause of vulnerability. After all, why should people necessarily respond to hazards using only their own abilities rather than supporting and being supported by those around them and societal services? Would we really wish to leave pregnant women, and others, to fend for themselves, or would we prefer to assist each other to avoid harm from hazards?

Other open questions remain. In societies where women have low status, could this play a role in women or girls being less likely to be helped or rescued than male relatives and friends? With these factors commonly permeating women's day-to-day decision-making, how much do they infuse aspects of vulnerability into their lives simply because of how society views and treats different genders?

Gender-based social roles and gender-based threats do not apply to women only. Men are also subject to physical and sexual violence, meaning that they, too, might avoid safe locations if they fear they could be attacked. Roles and expectations are foisted on men in many cultures, leading them to choose risky behaviours which increase their chance of being killed in disasters.

In the USA, more men than women tend to die in most types of inland floods. Possibly half of flood-related deaths in the USA are people trying to rescue someone else. Most trained rescuers, such as firefighters, are men, and most bystanders who take immediate action, driven by societal expectations, are men. A macho culture and the mentality that men should sacrifice themselves for women, as in 'women and children first' when ships sink, seem to increase the number of men dying in floods in the USA. Do these examples indicate that males have lower social status than women and hence are sacrificial in times of disaster?

Meanwhile, more than half of flash flood deaths in the USA occur in vehicles. The roles of drunk driving and of passengers goading their drivers to zip through swiftly flowing water are not known. Nor is the role of drivers desperate to be reunited with their families or to reach their children at school. Nor has the gender distribution among drivers and their passengers caught in vehicles in water, both survivors and fatalities, been thoroughly investigated. With many vehicles floating and being swept away in just centimetres of flowing water, the rule is to never drive, cycle, walk through, or otherwise enter floodwater. This rule is persistently broken for various reasons, sometimes with fatal consequences. And within all of these factors, how much is apparently instinctive and how much is socially and culturally constructed?

None of these points or examples, from South Asia to the USA, suggests that either men or women are inherently weaker or less intelligent than the other. Neither women nor men are inherently more afraid of water or less capable of making flood-related or tsunami-related decisions than the other. Instead, cultural roles foisted upon males and females create vulnerability, making it appear as if either men or women—depending on location and cultural context—must inevitably be worse off in disasters. These roles, experiences, vulnerabilities, and disaster-related consequences also need to be investigated for genders beyond the male–female binary. What we do know is that cultural roles are not generated by gender-differentiated capabilities and vulnerabilities; rather, gender-differentiated capabilities and vulnerabilities are created by cultural roles.

Notwithstanding the physiological differences between males and females, gender-differentiated vulnerability is largely manufactured through ideology. Notwithstanding the physical forces of water and their lethal potential, floodplains end up with flood disasters largely due to ideology. These ideologies are sometimes mainly in the political realm and sometimes intertwined inextricably with deep cultural, religious, and social traditions generating norms and expectations. The story of vulnerability is driven extensively by ideology.

Vulnerability by Economics?

Even when the desire to act is paramount before the wind blows, the earth shakes, or the water moves, economic factors might stop us doing what we want. The story of vulnerability includes how tackling vulnerability might be unaffordable or how decision-makers might state that tackling vulnerability is unaffordable. Whether by governments, businesses, organizations, or individuals, the requirement to spend now can be deemed to be too expensive, even while knowing that avoiding immediate costs for preventing a disaster may incur future costs which are far greater.

Certainly, individuals commonly do not have the money on hand that is needed to cover the measures they seek. During the

1995 heatwave in Chicago, elderly people perished from medical conditions related to heat and humidity, with the true cause being the lack of affordability of indoor climate control combined with the lack of choices available to them for affordable, safe, and cool locales. Any form of alternative accommodation to access air conditioning or a fan, whether a hotel or hostel, would have been out of the question due to cost. Even recognizing and accepting their own needs for coping with a heatwave in advance would not have spawned the options and resources they required to meet those needs.

Chicago now offers cooling centres with security during times of extreme heat as well as transport to and from these centres. Community centres, libraries, police stations, and other public buildings are opened up for people to get away from the sizzling sun cooking their homes. Overnight shelter beds might also be made available, although the centres are available mainly during the day which, by keeping people cool for several hours during the hottest part of the day, can suffice to keep them alive at home through sultry but slightly cooler nights.

Higher humidity makes the temperature feel higher and prevents sweat from evaporating, killing at lower air temperatures unless vulnerability is tackled. The length of time for which certain heat–humidity combinations are surpassed contributes to the danger level. If we broil during the day, but gain some relief during cooler nights, then being careful during the worst hours of the day could get us through a scorching period. When neither heat nor humidity relent overnight to the extent that our bodies can recover, the death rate skyrockets in the absence of reducing vulnerability. It can get to the point that fans assist little, since they end up simply blowing hot, humid air onto a suffering recipient.

We do not know with certainty what these lethal heat–humidity–time combinations are. Research continues to examine the demographic factors influencing when death is unavoidable and how precise any deadly thresholds or ranges are.

Likewise, periods of intense cold can kill if vulnerability is not addressed, so a number of cities offer public buildings as warming centres. Many deaths in homes during extreme temperatures, particularly intense cold, are blamed on 'fuel poverty', with residents unable to afford sufficient insulation or temperature control. As humidity augments the effect of heat, so wind speed augments the effect of cold, a phenomenon known as the 'wind chill factor'. Many residences block out wind, although they cannot do the same with humidity, yet poor maintenance or repair might enable bitterly cold draughts to enter. With heating and cooling centres, and transport to and from them, alongside support to tackle fuel poverty, planning and preparation can overcome vulnerability by economics to temperature extremes.

From Chicago to Adelaide, normal, seasonal weather which has happened for centuries ends up killing thousands annually because vulnerability removes options to stay alive in hot and humid or cold and windy weather. If all of us could afford adequate indoor temperature control, then we could make a choice about heating and cooling our rooms. The kicker is why large segments of the population have trouble providing themselves with such a basic necessity. Vulnerability by economics is more than just the fact of some families' lack of cash on hand each summer or winter. It dives into the realms of why energy suppliers might not lower prices to affordable levels for everyone; why residential buildings are not designed or retrofitted for the weather variations and trends in their location; and why those in rented accommodation are typically more fuel impoverished than owner-occupiers.

Interlinked social conditions crop up, from racism to inequity, as well as how much control governments should have over utility services, namely electricity and other energy suppliers. All topics converge on economic and governance structures denying people choices to reduce their vulnerabilities to drawn-out periods of intense humid heat or cold and wind. Without a range of choices, vulnerability forces people, as the saying goes, to freeze in the dark or to fry in the swelter.

These frightening plights showcase how economics, wealth, and poverty influence vulnerability. A general pattern, although not always met as there are plenty of counterexamples, is that poverty breeds vulnerability. Poverty provides us with fewer choices on where and how we live, so we end up occupying buildings which are constructed and maintained to worse standards than those in affluent districts. Poor areas might be denominated slums, favelas, shantytowns, ghettos, skid row, or informal settlements, which (with some justification at times) are perceived as being dangerous due to high crime rates. Those with resources and other options avoid them; investment and improvements are eschewed; and poverty perseveres. With fewer resources, poor people frequently have poorer nutrition, fewer opportunities for clean water and sanitation, and less access to adequate hygiene and healthcare, all conspiring to augment vulnerability and to deny options for dealing with it.

This description of a rich–poor gap leading to a vulnerability gap must be nuanced. Wealthier people sometimes knowingly or unknowingly select areas with multiple natural hazards for prestige, lifestyle, or environmental amenities including climate, because they can afford to. Affluent areas in Miami, Florida are concentrated along the Atlantic coast including the barrier islands, perfectly placed to be first in line for a hurricane's storm surge. Chic residences in San Francisco's Marina District succumbed to the 1989 Loma Prieta earthquake since they were built on reclaimed land which liquefied to mud during the shaking. Elsewhere in California, the walls and gates around ostentatious mansions consistently fail to stop wildfires transforming them into cinders—equally so with modest middle-class abodes and the small dwellings of the poor if fire-resistance measures have not been implemented. After the California firestorms of November 2018, celebrities posted photos of smoking ashes that were all that was left of their estates in Malibu and other areas in the vicinity of Los Angeles. They represented a small fraction of the total number of structures destroyed, mainly of poor and middle-class occupants, which itself paled beside the death toll above eighty. Celebrities have the resources to rebuild their mansions and to live well during the rebuilding, a luxury available to few others.

Lifestyle choices based on affluence permit humans and hazards to meet which otherwise would remain far apart. Heli-skiers and mountain climbers are swept away by avalanches far from any infrastructure. Flash floods knock out canyoners and forest campers flee wildfires. No especial principle rears up to ban people from making their own choices with their own resources, as long as others are not hurt. Instead, for annual holidays and day-to-day living, those who can afford potentially risky activities can make themselves fully aware of the hazards and vulnerabilities. They have the choices to take the necessary actions for reducing disastrous possibilities that may put others at risk as well as themselves. Not all opt to do so.

Those who can afford it can and often do choose to live in harm's way, for instance by purchasing a house on a floodplain to enjoy river views and access, or by a volcano's slopes for similar mountain amenities, without taking adequate measures to reduce hazard-related damage. Conversely, those who cannot afford it will understandably gravitate towards locations offering them livelihoods, no matter what the hazards. Soil enriched with flood sediment or volcanic ash can be especially tempting to use as farmland, so it brings people, infrastructure, and livelihoods into floodplains and volcanic hazard zones.

Villagers around Merapi volcano on Java, Indonesia, have developed their agricultural systems to use the ever-present volcanic activity. Ash fertilizes the areas where grass grows and hot gas emissions burn vegetation, supporting fast-growing grass over slow-growing trees. The villagers collect this grass to feed the cattle that form the basis of their livelihoods. Merapi's eruptions have led to casualties at least half a dozen times over the past century, yet the villagers' general view remains that the volcano's hazards are opportunities for them. Despite attempts at post-eruption resettlement, few have wished to move. Traditional beliefs and a sense of home contribute to reluctance to leave their land which, due to the volcano, provides them with a level of sustenance which they do not expect would be an option for them farther away.

The fundamental driver for vulnerability by economics, as with vulnerability by ideology, is choice. Wealthier people control vulnerability, their own and that of others, more than poorer people, because having greater resources means having more options for where to live and how to live. While ideologies, values, cultural traditions, and personal viewpoints lead to wishes that impact on vulnerabilities, more wealth increases the likelihood of fulfilling these wishes. And so we reach the conclusion that, as a general guide with exceptions, poverty begets vulnerability which causes disasters. Economic structures and interests then obscure this pattern, skewing our understandings of vulnerability.

Rankings of disaster impacts based on financial losses pervade the media and research journals. Reinsurance reports listing the year's top ten costliest disasters stamp the December headlines with figures of total insured and uninsured disaster losses. The USA brandishes its list of 'Weather and Climate Billion-Dollar Disasters' as a marker of the threats facing the country. Over a dozen of these disasters during 2018 killed at least 247 people, compared to the 28th September 2018 billion-dollar earthquake and tsunami in Sulawesi, Indonesia, which killed around eight times that number. The focus on financial losses appears to imply that the higher these losses are, the worse the disaster, yet such data do not fully reflect what people endure in disasters.

Those who are affluent enough to own two or more properties have much more economic value that can be destroyed in a disaster. Their properties might also be sited in disparate locations, potentially increasing the chance of one of them being impacted by a hazard. Having more to lose and more chance of losing something is certainly part of vulnerability, but loss is not just measured in absolute terms. If you are one of the few who own various half-million-euro properties and you lose one, then your losses are in six figures in euros. This amount is far more than you could ever imagine owning, let alone losing, if you are a subsistence farmer with three goats and two cows. Losing one of five animals, when that is almost all you have, affects a poor farmer far more than how the loss of one of five expensive buildings affects a property developer. If you can afford to own a few upscale properties, that by definition makes you less vulnerable than people who can afford to own only a few livestock. They, in turn, are better off than those who own nothing, renting their abode and tilling fields owned by others.

The element of choice worms its way back into the story of vulnerability by economics. People who can afford to own multiple properties should consider their financial situation for purchasing adequate insurance or, if insurance companies refuse to cover them, having enough cash on hand to deal with losses. They have enough resources to allow for backing up important documents, garnering hazard information, and evacuating themselves and important possessions when needed.

And why not examine changes to their infrastructure to reduce potential hazard-related damage? For the affluent, it is easy to assess the area for potential hazards, to consider possible damage to buildings, and to take measures to help the properties make it through a hazard. Perhaps windows should be replaced to withstand wildfires and appliances bolted down for earthquakes. If these changes are not affordable, then how affordable is having multiple properties—or death and injury in one of them?

In some cases, wealth might make little difference. A building on reclaimed land in an earthquake zone might tumble if the soil liquefies during tremors, whether the inhabitants are rich, poor, or in between. Survival for short time periods is possible inside a building enveloped by hot gas and ash blasting down a volcano's slope. It is hard to predict exactly how bad these flows might be and the exact survival rates for different flows—besides which the building might burn or collapse. For now, the best option is not to be there, no matter what our economic status, and affluence provides more options for being far away and for rebuilding afterwards.

These scenarios nevertheless show that not all vulnerability by economics is a one-way street from economic situation to vulnerability. Even where poverty leads to vulnerability, the converse of vulnerability leading to poverty emerges. Without any opportunity to address vulnerability over the longer term, poverty will be a cause and a consequence of disasters that are persistent or occur repeatedly.

Afghanistan must hardly enjoy its appellation as the 'Graveyard of Empires'. Past leaders who have marched into its lands include Alexander the Great and Genghis Khan. The British tried three times between 1838 and 1921 to enfold the country within its empire, efforts that resulted in horrific bloodshed and their own abject humiliation. Exactly the same consequences plagued the USSR's efforts to control the country from 1979 to 1989, one legacy from which was a setting covered in land mines. Civilians continued to suffer ordnance-related atrocities as the Taliban took over in the 1990s, adding further barbarities of oppression and persecution to the mix.

Following the 11th September 2001 terrorist attacks in the north-eastern USA, the al-Qaeda perpetrators were identified as being based in Afghanistan and backed by the Taliban's Afghan government. One more invasion force, this time from NATO, led in particular by the USA and the UK to support anti-Taliban militants, aimed to shape Afghanistan. The Taliban leaders were overthrown, retreating to the mountains with al-Qaeda's remnants. By the end of 2001, Afghanistan had gained plenty more explosives with the potential to harm civilians, along with a NATO-backed government implementing some traditional Afghan approaches to elections and governance.

Throughout these political and conflict stratagems, nature never desisted. Earthquakes, landslides, heatwaves, storms, avalanches, floods, droughts, lightning, and other natural hazards were simply part of life's routine in the landlocked territory. They were, in fact, as routine for ordinary Afghans as violence, despotism, combat, brutality, and exploitation. Dealing with nature has been straightforward for the country compared to dealing with what people, often based in far-away countries, have wrought.

Afghans have a long tradition of living in arid regions with large variations in rainfall. They would raise large herds of livestock, wandering the land to find water and eating some of their animals to get through tough times. Water harvesting has been standard, through snow captured and stored in wells called *cha* and through underground water transport and irrigation channels named *karez*. These mechanisms were undermined by the wars, destroyed or damaged in battles, while violence, fear of being attacked, and the risk of unexploded ordnance restricted movement with herds.

Millions of Afghans fled the country, languishing in refugee camps in Iran and Pakistan, or eked out a meagre existence elsewhere in the country as internally displaced people. Without hope, away from home, being treated badly by hosts who themselves struggle to make ends meet, and facing surroundings with munitions and conflict-ridden groups, displaced people find it difficult to deal with nature's typical phenomena such as earthquakes, extreme temperatures, and landslides. The hazards became disasters, necessitating aid on top of aid on top of aid. On 25th March 2002, just months after the fall of the Taliban, an earthquake flattened villages in northern Afghanistan. Over 2,000 people died, with many more left homeless. Amid the aftershocks, cold weather and rain leading to floods and mud interfered with aid assessment and delivery. In addition to having suffered decades of merciless war, the affected population cleaning up the earthquake rubble faced the prospect of explosives being triggered by the water and mud. In such cases, where do vulnerability–poverty interactions begin, and will they ever end by resolving both vulnerability and poverty?

Afghans are trying to do better for themselves and with international support. Suicide bombers, snipers, ex-combatants, competing perspectives of women's roles, internecine battles, and the killing of civilians by the warring forces are circumstances hardly making it easy to build the country's, families', and individuals' self-sufficiency for dealing with nature. Continuing disasters, most notably the human element of violence and oppression, leave little room for creating and maintaining momentum for the twinned long-term efforts towards vulnerability reduction and poverty reduction, so that each helps the other.

Vulnerability and poverty feed off each other through a lack of resources which prevents people from having choices that, if acted on, would help to reduce their vulnerability. The story of vulnerability to hazards encompasses vulnerability by economics, which is itself based on vulnerability by lack of choice.

CHAPTER 4

VULNERABILITY BY CHOICE

Vulnerability by Choice, but Whose Choice?

The Philippines is labelled 'one of the most disaster-prone countries in the world'. Poverty, armed conflicts, and discrimination are rife, setting up significant levels of vulnerability which are exposed by the variety of hazards affecting the country, including volcanic eruptions, earthquakes, typhoons, other storms, floods, landslides, and epidemics.

From this history, the Philippines has developed an impressive capability in countering the forces that create vulnerabilities. Several prominent groups lead the way in inspiring people to work with their neighbours and others around them to deal with hazards.[1] One technique honed in the country and then applied around the world brings together locals to build a scale model of their area using simple materials such as cardboard, yarn, and drawing pins.[2] The people then identify hazards, vulnerabilities, risks, and actions for reducing disaster potential.

Building these local skills has not been easy. The country reeled under the twenty-one-year US-backed repressive dictatorship of Ferdinand Marcos, who was overthrown in 1986. As the country was starting to recover from his corruption and greed, with the people rapidly learning that corruption and greed do not evaporate with democracy, the main island of Luzon was rattled by an earthquake on 16th July 1990. Little connection was made between this shaking and the island's dormant volcano Mount Pinatubo. Although the region around the mountain was known for its geothermal energy potential, the volcano's last major eruption had been several centuries ago, beyond the cultural memory of those running the Philippines, and Pinatubo was not high on the radar.

A sudden change in this view was demanded on 2nd April 1991, when numerous, small explosions of steam and mud from Pinatubo, perhaps triggered by underground molten rock set loose by the 1990 earthquake, led to a decision to evacuate 5,000 residents within ten kilometres of the summit. Two months later, tremors and explosive activity started ramping up, and on 7th to 9th June, the evacuation zone was expanded to within fifteen and then twenty kilometres of the summit. A US air force base sat within this zone and by 10th June, nearly 15,000 American personnel and most aircraft had left the base.

Two days later, Pinatubo exploded, producing an enormous ash cloud. The evacuation zone was extended by an additional ten kilometres, to thirty kilometres from the summit, displacing more than 33,000 people. Eruptions and seismic activity were now almost continuous, with 15th June providing the biggest eruption, in the midst of which a typhoon made landfall on the Philippines, soon passing just north of the thundering mountain.

Dozens of people died because they had taken shelter in buildings with roofs strong enough to withstand the ash fall, but which collapsed when the pulverized rock settling on the roofs absorbed the typhoon's water, adding weight. By 16th June, over 200,000 people had been moved out of the area around Mount Pinatubo, which was now quietening down. The huge amounts of ash deposited would, for years to come, mobilize as lethal flows called lahars, a Javanese word.

An indigenous group, the Aeta, with a population of up to 50,000, lived on and around the slopes of Mount Pinatubo prior to the eruption. The Aeta view the mountain as their saviour and protector; their duty is to live on the volcano. They were devastated by the eruption, because rather than persisting in giving them life and sustenance, their home exploded from under them. Following the eruption, the response and relief activities did not seem to have prepared for or demonstrated much interest in responding to the cultural values and needs of the Aeta. The Aeta were wondering if they, or perhaps outsiders in the form of loggers, miners, and geothermal energy exploiters, had angered the mountain. The destruction of their settlements and land meant that they could no longer be self-reliant, compounding their despondency.

They were forced into camps where easily preventable and treatable infections including measles broke out. Hundreds died from disease and/or malnutrition due to inadequate aid supplies. The government of the Philippines aimed to build new, permanent settlements for the Aeta without much consultation with the people. The Aeta resisted, thwarting these plans, so they were left dependent on external aid in temporary camps, whereas they would have preferred to be self-sufficient, providing their own shelter, food, water, and clothing. A few years after the eruption, many Aeta gave up and left the camps, returning home despite threats from lahars, earthquakes, and potentially more eruptions. At least they could provide for themselves on familiar territory, even if hazards might be lethal, a preferable situation to relying on others for potentially fatal assistance.

When caught between two terrible, potentially lethal choices, no wonder vulnerability is prevalent. Where is the choice here and who is choosing? If reducing vulnerability to and harm from disasters is desirable, then it is straightforward to make individual and collective decisions to achieve this—as long as the resources, choices, and opportunities exist. The Aeta did not have these because the available options were selected for them by outsiders without much consultation.

Similar situations can arise across diverse hazards, with epidemics as another example. During the Ebola epidemic disaster from 2014 to 2016 in West Africa, the most extensive in human history so far, people infected with the virus reached as far as the USA and the UK, yet outbreaks did not occur in those two countries. They had the knowledge, resources, and interest in averting the virus's spread, and they made the choice to do so, compared to countries such as Sierra Leone, Liberia, and Guinea, which required the world's help to arrest the epidemic because they lacked the options and resources to do so themselves. We have the abilities and assets to reduce vulnerability and to prevent a disaster, even with a virus as lethal and transmittable as Ebola. It is a choice—someone's choice—whether or not to do so.

The spread of Ebola in West Africa could have been impeded, if not entirely stopped, much earlier than it was. Aside from the failings of those countries' governments—due to a mishmash of causes ranging from inadequate post-war institutions and dominating external interests to corruption and incompetence—the international health monitoring and surveillance system, notably the UN's World Health Organization, was far too slow in responding.

In the years leading up to the Ebola epidemic, the World Health Organization was forced to cut its budget substantially, with the division responsible for preparing for and responding to epidemics being among those hardest hit.[3] Should the UN be blamed for the slow and inadequate response to Ebola? Or does responsibility rest with donating countries that lacked the foresight to want to prevent epidemic disasters? Who made the budget choices for the UN and why? Ebola and many other diseases are hypothesized to have afflicted human beings because we moved into new habitats, carving out space for ourselves among ecosystems harbouring the pathogens. As populations press on with moving into previously non-settled locations, often from greed but also through seeking self-sufficient livelihoods, should we aim to discourage this mobility, especially by providing better alternatives for living?

Or should we accept the expansion of human territory and be prepared for emerging or re-emerging diseases among the dangers from many other hazards? We all have so many different reasons for moving, due to so many different circumstances. Migration is a human characteristic and choice, as it always has been, leading us to new locations and potentially new hazards. Migration is not always a choice, though, when hazards force people to move.

In Ecuador in 1999, more than 16,000 people in and around the tourist town of Baños were evacuated just before the volcano Mount Tungurahua exploded, layering the area with ash and lahars. The choice was to leave or die. In Baños, the inhabitants were forced to leave, but as the length of their evacuation dragged on, as with the Aeta in the Philippines, many returned despite the risk of death. Similarly, as Mount St Helens moved towards its climactic eruption in May 1980, octogenarian Harry R. Truman refused to leave his home by the mountain's slopes. He became a local hero for disobeying the evacuation order, vowing to stay with his lodge and cats. His body was never found.

Sometimes, the reasons for moving from home are entirely voluntary, and don't involve consideration of hazards or vulnerabilities, just being for adventure, fun, and excitement. Young adults in New Zealand seek 'The big OE' or 'Overseas Experience'. They spend a period outside their country living, working, volunteering, and travelling as part of maturing and personal growth. Some end up settling abroad, building vibrant expatriate groups and bringing their knowledge to their new home. New Zealand displays many environmental hazards, including earthquakes, volcanic eruptions, landslides, floods, storms, tsunamis, avalanches, and extreme temperatures. New Zealanders on their OE should be aware of and ready for any such situations—assuming that they had taken on board this knowledge in New Zealand.

On the contrary, many are lulled into a false sense of security. Coming to London, UK, New Zealanders can adopt the same attitude as the locals that the city is safe because environmental hazards rarely happen. If you grew up learning how to react during an earthquake while living through numerous cyclone-type storms, such as in New Zealand's capital, would you be ready for anything in London? Would you presume the city to be much safer than your Wellington home? Or had you never been properly prepared even in Wellington? As a voluntary migrant to London, would you be more aware or less aware of the hazards, given that you are in a less familiar environment?

Not everyone has choices regarding migration into places experiencing hazards with which they are not familiar. As part of Norway's policies of supporting settlement around the entire country, asylum seekers are placed in all locations, from towns above the Arctic Circle such as Tromsø to still-snowy locales in the south such as Arendal. A Somalian might never before have encountered sub-zero temperatures, remaining highly vulnerable to winter hazards until they have learned the basics: driving through falling snow, walking on ice-covered pavements, and keeping pipes from freezing.

Most commonly, people move for a combination of seemingly forced and seemingly voluntary reasons. If we cannot get the education, job, or healthcare we wish for or need in our home location, our move is partly forced (what we need) and partly voluntary (what we wish for). Nonetheless, we have some level of choice in terms of getting to know our new home and the hazards and vulnerabilities that come with it.

After an earthquake disaster, most people would prefer to stay near their property and land to rebuild structures, livelihoods, and networks. But conditions do not always permit this option, whether due to safety, absent services, or lack of support to rebuild fully at the preferred pace of those affected by the disaster. Their decision to migrate, temporarily or permanently (or temporarily morphing into permanently), is both voluntary and forced. It is voluntary migration because the people (that is, those who have the resources) make a choice. It is forced migration because it is dangerous and counterproductive to camp where systems and services—health, food, water, energy, and education—are inadequate.

Yet these systems and services are regularly absent for migrants if they do not know how to access them in their new location, or if they are unwelcome and marginalized by their hosts. Migration by itself does not inherently increase or decrease vulnerability to disasters, because it depends on the circumstances and how the migration is managed. If we are strongly attached to our home, our land, and our way of life, then shifting location is devastating, unhealthy, and insecure. We might never be able to adjust to the move, and our vulnerability increases. If we are ecstatic about migrating, embracing all that is different about our new place, then we learn, adapt, contribute, and enjoy, so our vulnerability decreases. Our enthusiasm and spirit could even reduce the vulnerability of those around us.

Some of us cannot take full responsibility for reducing our own vulnerability through mobility, because it is under the control of others. Prisoners are an example. The point is not that prisons should be abolished or that convicts do not deserve to have their movements controlled—although many places abuse and violate the justice system. Even where prisoners are rightfully incarcerated, their punishment was jail, not death in a disaster. During the 26th December 2004 tsunami, hundreds of Acehnese prisoners died because they could not escape from their cells as the water rushed in.

Individuals' responses play only part of the role in determining how migration affects vulnerability to disasters. If we are forced into makeshift camps with poor food and water supplies and limited hygiene, as the Aeta were after Mount Pinatubo's eruption and as many Haitians were after the 2010 earthquake, then our vulnerability increases.

Following the 1994 genocide in Rwanda, refugee camps were set up just across the border in Goma, in what was then called Zaïre and is now the Democratic Republic of Congo. Aside from the complication that some of the refugees were accused of being genocide perpetrators, and ran the camps as dictators of their demesne, many of the shelters were placed on old, hardened lava flows from the nearby volcano, Mount Nyiragongo, because it was the main land area available near services. In January 2002, during a rebellion against the national government far away in the capital Kinshasa, the volcano erupted, sending streams of lava through the Goma camps and city. The displaced became re-displaced, with few resources available to deal with the volcanic hazards amid a violent conflict besetting those already forced to run from their country as refugees, remembering that some of those refugees were implicated in generating the conflict that had made them such in the first place. The intertwining of conflict, hazards, and migration becomes complicated.

Under international and national laws, countries have the right to control cross-border movements. The consequence is that some nationalities have easier movement around the world than others. This is not to judge the right of a democratic parliament to set boundaries on who is permitted to enter and leave their country. But it demonstrates how vulnerability can be created through choices to control mobility (which may have defensible reasons), meaning that people lack choices to cross an international border in the wake of a major hazard.

Permitting international mobility can potentially augment vulnerability too. If we voluntarily elect to travel, should we take responsibility for our own vulnerability or should we expect our destination to provide what we need? The answer is both. Neither individuals nor organizations nor governments can or should accept full responsibility for everyone affected by disasters. These actions are a collective responsibility and choice, meaning that we must work together, all contributing what we can within our abilities and resources.

Some of these abilities and resources relate to communication, with our skills differing from day to day. Hunger, lack of sleep, or using substances which alter our mental state—such as alcohol, cannabis, or sleeping pills—influence our awareness, how we express ourselves, what we perceive, and how we interpret and understand the world around us. Imagine being under nitrous oxide (laughing gas) for wisdom teeth removal or general anaesthetic for a biopsy when an earthquake rumbles through or a flash flood warning requires immediate evacuation. These scenarios point to a choice to site all health facilities in safe buildings in safe locations with on-site utility backups.

Such measures cannot cover all variations in communication. Expression and interpretation abilities for literacy, oral expression, body language, sign language, and other forms of communication are affected by age, educational background, different personality types such as emotional or analytical, state of health, professional background, and cultural background. These characteristics interact with and influence each other. Some are a matter of individual choice, some arise from others' choices, some are forced by physical circumstances, and some involve combinations of these factors.

Communication systems require resources. Some people cannot afford a mobile phone or television, while hearing aids cost money, as do their batteries. If a person is already marginalized, perhaps by not speaking the dominant language, then they might not have the knowledge, connections, or time to examine different hearing aids or warning services to pick the best for themselves. Cultural and educational backgrounds affect communication abilities if provisions are not made for everyone.

When, like New Zealanders on their OE, we travel to places where we are not aware of the hazards and how to deal with them, or where we do not know the languages or cultures, we can become highly vulnerable if the authorities and locals do not factor in our needs. As visitors to other countries, we might not know any of the local languages and we might have a limited understanding of local customs and body language. This is a major predicament in tackling vulnerability.

If we do not speak, read, or understand any Chinese dialects, should that stop us from holidaying or closing a business deal in Shanghai? Should we expect all visitors to Edinburgh to speak English or Gaelic, and all visitors to Ottawa to speak English or French? This means having the vocabulary and grammar as well as being aware of the body language and culture. In any case, in many parts of Edinburgh, Gaelic would not be especially helpful, just as in some parts of the Ottawa region, only English or only French would be needed.

Many expatriates choose to have little knowledge of local languages and cultures. Others decide to learn as much of these as they can and as quickly as they can. It always takes time, and a language and culture learned as an adult might never be at the same level as growing up with it. People learning a language or culture later in life can miss nuances in warnings, they may not have previously come across an idiomatic expression, or they may be unfamiliar with technical terms. Even native speakers can be confused by dialects, local expressions, and jargon.

In India, a *ghat* or *ghaut* is a set of steps leading down to a river used for bathing or, sometimes, a mountain pass. In some of the smaller, anglophone islands of the Caribbean, a *ghaut* is a steep ravine with a small river leading to the ocean. It can become powerful, fast flowing, and deadly in the wake of a hurricane, sweeping away vehicles and people foolhardy enough to try to get across. Speaking English from an Indian perspective in the Caribbean might not prepare everyone for the local post-hurricane context of *ghaut*.

Even the word 'hurricane' is regional. When a rotating, large-scale storm within certain latitudes reaches a specific wind speed, it is called a 'hurricane', a 'cyclone', or a 'typhoon' depending on the ocean and longitude where it originated. This seems arbitrary, and it is. Specialists know the differences and similarities among extreme storms and they use their vocabulary for effective communication among themselves. Their professional background and the shared technical terms of a field support them in pinpointing exactly the form of hazard being considered. For those outside the field, the story is not the same. To respond to a warning, they are generally not so concerned about whether an environmental phenomenon is technically termed a tornado, downburst, microburst, waterspout, landspout, funnel cloud, fair weather vortex, wind shear, gustnado, or dust devil. Nor would the various meteorological origins and characteristics of these phenomena be of primary interest. An aircraft's pilot needs to know which are dangerous, how to avoid them, and how to react if caught in one. Anyone driving a vehicle, at home, at work, or elsewhere, needs similar information for similar choices.

Language for hazards, vulnerabilities, and disasters can be used to confuse rather than to assist. If I explain that a pyroclastic density current from a dome collapse or column collapse typically has a velocity over 22 m/s and a temperature over 473 K, then it will be mainly volcano experts who glean the implications. If, instead, I describe how some volcanoes can produce an ash-gas cloud hotter than boiling water that moves faster than motorway speed limits as it flows down the mountainside, then hopefully most audiences would be interested in knowing how to avoid being killed by it.

Inadvertent misunderstandings have led to lethal choices. In 1984, the volcano Nevado del Ruiz in Colombia appeared to be awakening. Scientists closely monitored its activity into 1985 and feared an eruption, carefully mapping out possible hazards and scenarios. They were confident they knew who was in danger from the expected lahars as the eruption's heat melted snow and ice which would mix with the ash. At the time, the word 'lahar' was not well known in Spanish, so to communicate with the public, 'lahar' was translated into *avalancha* (meaning 'avalanche'). The locals quite sensibly reacted by thinking that Nevado del Ruiz could not produce avalanches which would reach their village, so why worry? Many made a risk-informed choice not to evacuate.

On 13th November 1985, the volcano erupted as expected; churning, debris-filled lahars raced down into the valleys as expected; and warnings were issued as expected. Over 23,000 people were buried alive. Most of the fatalities were in Armero, built directly atop lahar deposits from Nevado del Ruiz's 1845 eruption, which had killed hundreds in the same location. Effective communication and appropriate choices based on communication can create the difference between an annoying hazard and a devastating disaster.

The ability to communicate is not always moulded by individuals' own choices. Some people are born deaf or unintentionally and uncontrollably acquire hearing disabilities throughout life. A tornado or tsunami warning system involving only sirens or loudspeakers cannot serve everyone in the population.

As we have seen, people with disabilities are often assumed to be especially vulnerable to disasters, to lack choices for reducing their vulnerabilities, and to be in need of extra help. But this happens only if society decides not to consider everyone's needs as being similarly important,[4] such as having only sirens or loudspeakers rather than a variety of media for warning messages. A cyclone or tornado shelter with only stairs for entry could never meet everyone's needs. People who use wheelchairs or who have just had a hip replacement would need to be carried into and out of the shelter. Consequently, the shelter cannot be fully effective.

It is the same for evacuation, as powerfully explained by Marcie Roth, who has dedicated her career in the USA to helping people with disabilities deal with disasters. Marcie tells the story of Benilda Caixeta who, as a quadriplegic in New Orleans' Upper Ninth Ward, had been trying to evacuate from hurricane Katrina at the end of August 2005. Evacuation planning for people with mobility constraints had been poor. The emergency services were unable to help and Benilda had made numerous unmet requests to the local accessible transport system. Marcie explains how she tried to use her contacts and experience to assist, continually assuring Benilda over the phone that help would soon be coming to her. As they were talking, Benilda suddenly yelled that water was entering her house and then the phone line cut out. Five days afterwards, Benilda's body and her wheelchair were recovered.

Despite knowing the hazard, vulnerability, risk, and impending disaster, one more funeral was needed because disaster-related knowledge was not applied long before Katrina was on the horizon. Continuing, deep-rooted inertia and apathy of the political system, far beyond what either Benilda or Marcie could ever tackle, doomed Benilda no matter what the hazard, because choices were made not to address known vulnerabilities over the long term.

Tackling vulnerability means making the choice to meet everyone's needs. If a shelter has only stairs or ladders, then people without the ability to use these entry and exit points are not inherently more vulnerable in disasters because they need help. It is the choice of shelter design and construction that has made them vulnerable, because now they must request extra assistance and hope that someone is around who is willing and able to provide support.

Building a ramp for each shelter might sometimes cost more upfront and require more space. Providing extra assistance to people with mobility disabilities during an evacuation might cost more at the time and might force some people to rely on others. But permitting deaths in disasters, such as Benilda's, has incalculable costs. The choice is ours regarding which costs we accept. To reduce all vulnerability over the long term, and not simply respond during a disaster with technical means, people must have choices for supporting themselves in terms of information, decision-making, evacuation, and shelter. Dignity and self-help are part of the human requirement for avoiding disasters by reducing vulnerability and supporting personal independence and autonomy.

Marcos Eduardo Barquero Varela, who lives in San José, Costa Rica, has visual disabilities and occasional seizures. He describes how his grandfather instilled earthquake safety in him with the advice to remain calm and not to run. Marcos applied this guidance on 5th September 2012, when a major earthquake shook the country. The house withstood the shaking and Marcos was able to quieten down his maid who wanted to dash into the street where, Marcos was afraid, electricity wires could pose a danger or running could cause a person to slip and fall. Because he had an earthquake-resistant house and had been given sound instructions before the hazard which he had listened to, Marcos' disabilities were irrelevant to his survival in the earthquake. And he was able to choose to help the maid, who does not have disabilities. Vulnerability reduction long before the earthquake meant that Marcos retained his dignity, applied self-help, and assisted someone else who should have been taught beforehand to help themselves.

Are similar choices available for everyone? Many people rely on oxygen and others require twenty-four-hour care, while those with certain cognitive and intellectual disabilities will have specific requirements for communication, evacuation, and shelter. People with compromised immune systems might be highly vulnerable to infection within the close confines of a shelter. Why should society make choices and incur the costs of making arrangements under all hazard circumstances to reduce vulnerabilities for all these people?

The answer surely lies in the final word of that last sentence: because they are people. They have the same rights as everyone else, they deserve the same dignity as everyone else, and they should be given the same opportunities as everyone else to deal with disasters.

There is no question that with adequate resources, and appropriate attitudes and preparations, people with a huge range of capabilities covering all states of health and communication abilities would have choices about tackling their own vulnerabilities. They would rarely need assistance for evacuating, reaching shelter, and all the other components of dealing with disasters.

Any 'burden' would reveal itself only in circumstances in which these elements providing people with choices are not in place—which unfortunately is typically the norm.

There are always choices to reduce vulnerabilities. The question, 'whose choice?', matters most. Those affected most by vulnerabilities generally have the fewest choices available for tackling them.

No Vulnerability is Inevitable

For the vast majority of the world's population, full resources, choices, and opportunities are not available to tackle their disaster vulnerability. Others make decisions which impose vulnerability on the majority and stop them from dealing with it.

Resolving this situation involves identifying the people, groups, politics, power games, and social structures responsible for decisions that create disaster vulnerability. Sometimes, legal recourse is sought to establish blame, to enforce action, and to expand choices. Sometimes, the legal system itself creates vulnerability to disasters. Questions of ethics and ideology emerge, because many groups with power do not, when it comes down to it, wish to reduce the vulnerability of other groups to disasters. They create and perpetuate systems and structures which form and sustain disaster vulnerability deliberately or as a by-product of their choices. They perceive that they gain advantage from vulnerability, as well as from disasters—and this can be true in the short term—so that they have little incentive to modify their ways.

Ultimately, vulnerability emerges from political decisions and for political reasons. If we prefer the pathway of reducing vulnerability, then we need to discover and apply methods of facing those who create vulnerability and the processes they use to do so.[5] Espousing values and engaging in actions which reduce vulnerability within any social system require conscious choices. Because we humans create vulnerability, sometimes deliberately and sometimes unwittingly, there is much we can do to reduce it. But we must actively choose to do so.

The factors generating vulnerability described in this and the previous chapter, and other factors not addressed in this book, cover physical aspects of individuals (such as age, gender, and state of health) and aspects of society (demography, economics, cultural norms, and the built environment). There is no neat link to outcomes, such that a specific trait or situation always increases or decreases vulnerability. The reasons why vulnerabilities are not overcome and why advantages are not reaped relate to ideologies, governance, prejudices, assumptions, resources, power relations, and availability of choices. In most circumstances, certain groups within society make choices for other groups, making or breaking their vulnerability. Individual and societal characteristics then appear to create vulnerability, yielding a false impression of its underlying causes. This does not remove individual responsibility in cases where people have choices and resources yet still select vulnerability, sometimes for defensible reasons. But it does mean recognizing that vulnerability arises from a balance between options for oneself and imposing options on others. Delving into the nature and causes of vulnerability makes it clear that people are not inherently vulnerable to hazards, but are made vulnerable by society.

CHAPTER 5

MAKING THE CHOICE

Controlling Hazards or
Controlling Vulnerabilities?

The Sahel, depending on how it is defined, traverses over a dozen African countries in a band stretching 1,000 kilometres north to south and more than five times that length east to west. Bordering the southern edge of the Sahara Desert from the Red Sea to the Atlantic Ocean, the semi-arid landscape of the Sahel is prone to dust storms and dry spells. To live in this environment, the peoples of the region have combined nomadic and farming lifestyles, which have helped to avoid overtaxing the soil through continuous agriculture and which have contributed to managing periodic and sometimes lengthy periods of drought.

These societies have been far from idyllic. Even when doing reasonably well at coping with environmental changes over millennia, many of the peoples have practised child marriage and female genital mutilation. These forms of internal oppression and violence create as much disaster vulnerability as external oppression and violence through colonialism, war, corrupt governments, and racism. Yet the Sahel's inhabitants have plenty to teach the rest of us about living with fluctuations in rainfall.

Why do periods of low precipitation today and in recent decades seem to have far more adverse impacts than previously? Unlike the past, contemporary droughts are not so much about a drop in the amount of water falling from the sky. Today's droughts do not always originate in nature. Instead, the disasters are caused by long-term management—or rather mismanagement—of water and people.[1] Overpopulation is a factor, as is overgrazing, each stemming from forced changes in interactions between humans and the environment across the Sahel.

Each individual, human or animal, requires a minimum amount of water to survive, and if this minimum level is not available, then deadly shortages manifest. Stabilizing human and livestock populations, to ensure that the water available is sufficient for them, always leaving extra water in case of prolonged periods of nature-related drought, are evident steps towards averting disasters. But stabilization of populations requires a suite of efforts. For livestock, it means regulations, and their adequate monitoring and enforcement. For people, it means options for affordable contraception, family planning, reproductive choices for women and men, and infant and parent healthcare. These measures are a long way from rainfall and lack thereof, but they persist at the core of dealing with drought disasters.

Another major factor in the overuse of water in the Sahel is the loss of options to maintain a nomadic lifestyle. European empire-building and the devastation of cultures and societies from colonialism were followed by independence and devastation from post-colonialism, often as a result of national leaders emulating colonial or pre-colonial ways. Many presidents and prime ministers, and those opposing them, around the Sahel proved to be as violent, corrupt, and abusive as their former colonial masters and their pre-colonial local leaders. Many recent and current leaders were trained in their colonial countries and were supported by, or were puppets of, the former colonial powers and global superpowers.

The new international borders restricted nomadism, severing movement-related lifelines which had helped the people to address water shortfalls previously. Forced settlement to end pastoralism increased local water use and undermined proven ways of living through times of less water. It is a similar situation with food. Traditionally, some food would be preserved and stored so that it would be available during difficult times. Food would commonly be exchanged and shared if one group was in trouble. Rather than settling on a single large farm, many nomads owned small plots in different locations, so that if one place or ecosystem was struggling, then the others would support them.

The move to contemporary ways of life, especially forced settlement, yields efficiencies in the short term. Services are always nearby, populations can be identified and counted, and logistics for goods are simplified. Land protection laws, hunting restrictions, and ecosystem conservation have strong benefits and can end destructive traditional practices. But ending pastoralism can create longer-term vulnerability if the disadvantages of nomadism are overcome without retaining the advantages. Knowledge of the wider environment is gleaned through continual mobility. Resources can be spread out, so the loss of one cache, herd, or plot of land does not mean starvation. A balance should always be sought to meld the best of external approaches with the best of what the people know already.

The Sahel is not the only area to suffer. California and Taiwan endured drought disasters in 2015. Some commentators buried the long-standing vulnerabilities beneath blame on the lack of rain, which was inevitably attributed to climate change. While climate change may exacerbate some situations, other analysts pointed out how poorly water had been managed. In California, industry, agriculture, and population expansion suck aquifers and waterways dry. Parts of California are arid, but the residents insist on maintaining leafy suburbs graced with trees from far wetter climes. Unsurprisingly, overusing water in a desert leads to drought. Across the Pacific, a plethora of wanton practices can be identified in Taiwan. Its pipe system leaks; many reservoirs suffer from silt, which reduces their capacity; and wasteful usage patterns are evident throughout the country.

It should be no revelation that a dip in rainfall leads to water shortages, particularly when people do not alter their water consumption patterns. Part of the difficulty in changing people's behaviour comes from the variation in rainfall. As we have seen before, the frequency of a hazard has implications for vulnerability.

What if episodes of less rainfall recur every few years? Societies such as those of the Sahel have incorporated this hazard into local knowledge, preparing for less rainfall through day-to-day and decade-to-decade living. Such awareness is engrained into their lives, becoming incorporated into their planning, rather than being experienced as a shocking extreme provoking a panicked response. Just as in the case of living by a river that floods regularly, people act to reduce their vulnerabilities, and less rainfall no longer becomes hazardous. In contrast, if drought rarely occurs, then it may not be factored into local knowledge about life and livelihood. We might not think about less rainfall and we will not be ready for it. Our vulnerability is high, with any drought being hazardous. Awareness and planning for potential environmental events and processes, through tackling vulnerability, can determine what is and is not a disaster.

There are many successes—traditional and modern (and much can be achieved by amalgamating the two)—from which we can learn and which should inspire us. Yet the numerous positive examples are not at present outpacing the creation of disaster risk. We can turn this situation around by learning from each other and formulating new approaches to move towards reducing disaster vulnerability.

These efforts never preclude ways of controlling hazards that avoid creating vulnerability. We have already seen examples of hazard control which create vulnerabilities, from wildfires in Australia to flood barrages in London, rather than choosing to work with nature constructively to help both humanity and the environment. What about other hazards? What possibilities, which also reduce vulnerability, exist for controlling or eliminating hazards? We need to be careful in assuming that we can make this work, since we are not always aware of the full consequences of our actions.

When it comes to biological hazards, humans have been remarkably 'successful' at eliminating 'hazards' posed by many species by driving them to extinction. Reckless destruction of ecosystems, viciously efficient hunting, and the transport and introduction of species into ecosystems where they may lack predators or other natural checks on their populations have decimated flora and fauna around the world. In so doing, we harm ourselves by undermining livelihoods, wiping out the advantages which nature brings, and setting in motion deeply interconnected processes involving air, water, land, and other species.

In some cases, the biological hazard may also have beneficial aspects for us. Snake venom contributes to drugs combatting hypertension. The deadly nightshade plant is poisonous to humans but it is used to detoxify certain pollutants. Large animals such as bears might be dangerous at times (although they often attack people only to defend themselves or their families) yet many cultures need them to provide meat, hide, or fur.

Some of the most hazardous and most lethal species have been deliberately targeted for the benefit of humanity: microorganisms that cause disease. Quantifying the lives saved from the deliberate eradication of smallpox and rinderpest is not an easy task. The death toll was, at times, millions in a smallpox epidemic, from the Roman Empire to the European invasion and colonization of North and South America. Estimates of rinderpest deaths run into hundreds of millions of cattle alongside a devastating toll on wildlife, knocking out food supplies and livelihoods, with a subsequently enormous tally of human casualties through starvation and the undermining of pastoralist cultures.

Many other diseases caused by small organisms, from bacteria to parasites, lead to horrible suffering daily, in addition to the epidemics and pandemics which hit the headlines. Those targeted for eradication, which means getting rid of the hazard, include polio, measles, rubella, and dracunculiasis, also known as Guinea worm disease.

Local and regional results demonstrate what is feasible. Polio and dracunculiasis currently exist in very few places—and these locations are marred by conflict which impedes effective disease eradication. Polio, targeted in 1988, was soon eliminated from most areas of the world. It remains entrenched in Afghanistan and Pakistan because of conflicts, in which polio vaccination workers are also murdered. It reappeared in Syria in 2013 after a hiatus of more than a decade, because the war wracking the country inhibited vaccination and created conditions that favour the spread of the disease.

In the case of dracunculiasis, the worm grows in people who ingest drinking water contaminated with small crustaceans infected by worm larvae. The worm causes a blister which ruptures, so the infected person soothes their pain by immersion in water, into which the next generation of larvae are released, to be ingested by the crustaceans. The campaign to eradicate the disease was founded in the early 1980s. It focused on providing safe drinking water, while working with infected people to ensure they cool their blisters and rid themselves of the parasite without reintroducing larvae into water supplies. By 1990, the efforts had paid off: dracunculiasis was prevalent in only twenty countries. By mid-2015, the list had been reduced to Chad, Ethiopia, Mali, and South Sudan, all of which had violent conflicts that were inhibiting the eradication.

Throughout history, many other local eradications of microorganisms and parasites have saved lives and substantially increased quality of life. Following separate cholera epidemics in Paris and London in the nineteenth century, the cities constructed their first sewage systems, preventing further outbreaks. Malaria used to be prevalent in the southern US while the parallel affliction of ague felled populations in the marshes of southern England. Dedicated efforts rid those areas of the hazards, though in some cases the methods used led to other problems. In England, the coasts were transformed by draining marshland, filling in land, and building up the shorelines to create the huge vulnerability to storm surge flooding witnessed today, such as for Canvey

Island. The hazard of ague was removed, but at the cost of creating vulnerability to floods. Future efforts to reduce flood risk should ensure that other hazards and vulnerabilities are not augmented.

Exactly the same arguments apply to other hazards which humanity aims to control. Reservoirs are used to control flood and drought hazards, with claims and counterclaims about their relationship to earthquakes. Did the weight of water on fault lines under the Colorado River's artificial Lake Mead along the Arizona–Nevada border really lead to increased seismicity in the 1930s as the reservoir filled? It is hard to know since seismic records prior to this decade are poor. How accurate are the assertions that filling the Koyna Reservoir near Mumbai, India in 1962 led to a 1967 earthquake which killed around 200 people? Observations seem to show a correlation between the reservoir's water level and earthquakes in the area, but the presence of a correlation does not necessarily mean direct causation.

No matter what the answers to these questions and no matter what choices could be made to control the interlinked water, drought, and earthquake hazards, choices could have been made to control vulnerabilities. This is particularly the case since neither Lake Mead nor Mumbai are necessarily immune to earthquakes otherwise. The earthquake history of Las Vegas, the largest city near Lake Mead, and the surrounding area remains a rich topic for detailed investigation, especially to consider what knowledge from past centuries the indigenous peoples of the region could offer. Records of seismicity around Mumbai reveal likely major earthquakes in 1594, 1618, and 1678, in addition to many recorded tremors in the nineteenth and twentieth centuries. The real headline is that this area's geology suggests large earthquakes recurring on the millennial scale, so more recent building collapses may not give us much evidence as to what could happen in the future.

Whatever the uncertainties concerning the seismic hazards, the vulnerabilities are long-standing and are well known. From the gaudy high-rise architecture of Las Vegas' gambling heaven/hell to the infill on which much of present-day Mumbai is constructed— and which is prone to liquefaction in earthquakes—we certainly know that not all buildings will remain untouched after the next big one in either location. One calculation of earthquake impacts for Las Vegas suggests hundreds of fatalities and thousands of injuries.[2] Given the population, the infrastructure, and the land on which it is built, Mumbai could suffer casualties which are an order of magnitude or two higher.

Creating or augmenting hazards, such as earthquakes from reservoirs or fracking,[3] cannot be justified on the premise that reducing vulnerability will take care of any problems. If we alter nature or manipulate the environment, the potential consequences need to be considered, accepted, and addressed. Hoping that it will all work out for the best and no harm will likely be done is not enough, especially when our understanding of the hazard and abilities regarding hazard modification are far from complete and full of uncertainties and unknowns.

Earthquake modification epitomizes this point. Numerous schemes have been optimistically proposed for controlling tectonic shifts; perhaps by building reservoirs over faults, or setting off small explosions, we can induce small earthquakes, thereby averting a large one. Perhaps injecting water into faults will lubricate them to avoid a sudden, large slip. These examples represent exciting science, provided that they are confined to the computer, the lab, and theoretical discussions rather than messing around with real faults, particularly in places with high earthquake vulnerability. Considering how straightforward it is to control (and preferably reduce) earthquake vulnerability, this is the pathway of action needed, rather than toying with the controversies and vast voids in current knowledge regarding earthquake causes, patterns, and control.

Cloud seeding for controlling weather is just as divisive. Experiments around the world have hoped to suppress hail, increase precipitation during dry periods, disperse fog, and modify the intensity and track of tropical cyclones. The process basically entails scattering small particles, such as dry ice or silver iodide, into a cloud with the idea that liquid or solid water from the cloud will coalesce around the particles until they are heavy enough to drop as rain or snow. Despite decades of cloud-seeding activity, no verdict has been reached regarding if, how, and under what circumstances cloud seeding succeeds or fails.

In 1947, the first documented effort to seed a hurricane took place as a storm passed Florida and Georgia, moving north-east. Following seeding, the storm obediently switched direction and slammed into the coastline of Georgia and South Carolina, prompting talk of compensation claims for the damage. But proving that human intervention led to the hurricane's observed behaviour is fraught with difficulty. What if it were shown that seeding reduced the hurricane's severity even as it changed its track? What about responsibility for choosing to not seed a hurricane which is clearly going to be destructive? Nevertheless, fears of litigation led to many of the observations from 1947 being classified, as well as the discontinuation of some weather modification programmes.

More to the point, should the people who built and maintained the battered infrastructure be accountable for some or all of the hurricane damage? This group does not necessarily encompass the people living there at the time, since they might not have been given the opportunity to acquire hurricane-related knowledge or the power to act on knowledge they had. The responsibility, perhaps, lies with those who could have highlighted that previous hurricanes had made landfall in these locations and who had the power to request changes on the basis of this knowledge. Whether or not cloud seeding made any difference to the hurricane of 1947, a storm could have hit the same places with the same, or worse, wind and water hazards. The damage was not the fault of the hazard—the hurricane—or of factors modifying or controlling the hazard. The damage and disruption, in other words the disaster was, once again, rooted in vulnerabilities, and the human decisions involved in creating, addressing, and controlling (or not addressing or controlling) them.

What about Climate Change?

Climate change infuses many discussions about disasters today. Understanding the basics of the science is essential for understanding the effects of climate change on specific disasters—or lack thereof. Climate, by definition, is average weather. We record temperature, rain, wind, and other weather features; we average them over several decades, typically thirty years; and the resulting numbers define the climate.

According to UN definitions, climate change is literally a change in climate, namely how weather statistics shift.[4] Some calculations focus on the shifts attributable to human influence only, rather than including natural trends and variations. There is no question that the impacts of human activity substantially affect the climate and that we are now witnessing the effects through swiftly changing weather regimes. The weather and climate have, of course, changed since the earth formed. Today, we see large-scale, rapid changes due to significant human influence, yet the definition of 'climate' and these statements about climate change show that they are about hazards only. Weather can be hazardous; climate, or average weather, describes hazard statistics; and so climate change is mainly about how weather hazards vary.

But disasters are caused by vulnerabilities, not hazards! It seems that neither climate nor climate change, by definition, can be or can cause a disaster. For a disaster to occur, vulnerabilities must be present, and they inevitably are. Even where climate change makes hazards worse, or increases the likelihood of extremes, in most cases we could choose to reduce vulnerability to avert disasters.

In the Sahel, modern political structures tend to preclude tackling all drought vulnerabilities without causing further troubles. Choices exist, but the system makes most of them unpalatable to those with the power to make the choices. Promoting pastoralism or softening international borders to help people deal with drought is not on today's political agenda for the Sahel. These political roadblocks are about people, not climate change. If we fail to plan for and work within the context of a changing environment, including climate, then disasters can and do happen.

As the air warms due to climate change, heatwaves and precipitation-related droughts become more likely. We do not fully cope with today's heat, droughts, and wildfires, while vulnerabilities are increasing and we know that without tackling vulnerability, worsening hazards can mean more disasters involving heat, droughts, and wildfires. Warmer air also holds more moisture, meaning more intense rainfalls. Again, without addressing storm and flood vulnerabilities, more and worse flood disasters can be expected under climate change. But not all aspects of storms will be worse. Projections under climate change for hurricanes, typhoons, and cyclones suggest that fewer will form. When one does form, the warmer air and ocean will augment its intensity. Additionally, it seems that many storms are moving more slowly along their tracks, leaving more time to dump rain in one place. Climate change is expected to lead generally to fewer storms, but they are likely to be more powerful and dramatic.

Increased atmospheric moisture means more intense humidity, and as we have seen, heat and humidity, especially for a few days in a row, can be lethal. In the Sahel, many people live in open dwellings without air conditioning and work outdoors for subsistence agriculture. Higher heat and humidity will curtail their performance, while people and animals will require more water to stay healthy, increasing the likelihood of drought. Crops, too, will suffer as evaporation rates increase, so we might need to substitute present ones with plants having different irrigation and tilling requirements. When heat–humidity combinations beyond human tolerance are reached, then outdoor work must stop, rather than simply going slower.

Higher temperatures make it harder to learn at school and to do homework at night, impeding education. Adding air conditioning or using fans boosts expenses and energy consumption, although plenty can be done for cooling building interiors with traditional and modern architectural practices that encourage natural ventilation. Due to climate change, thresholds of heat and humidity might be reached which surpass the abilities of natural cooling techniques.

Ultimately, without action on climate change, it looks as if many large swathes of land will end up too hot for working or learning during the day, and it will not always cool down at night. With a heat–humidity combination beyond human survivability lasting successive days and nights, children and the elderly will start dying because their bodies have less capability to regulate their internal temperature. Heatwaves in the future could be disasters caused by climate change for which we have little comprehensive opportunity for reducing vulnerability. As such, they end up being an exception to the argument that climate change does not in itself cause human disasters.

Nomadism would help by enabling people to move to cooler locations, but the Sahel exemplifies the barriers which have been created. Elsewhere, from Sydney to Ahmedabad, nomadism might not be an option, since it is not necessarily part of the culture and not everyone wishes to or can afford to move continually. In the end, changes to average weather are likely to create close-to-intolerable conditions for living and working for stretches of days at a time. No one should wish this form of weather on anyone. But we can deal with it. We can redesign dwellings and offices to augment indoor climate control, natural or mechanical; we can mechanize outdoor work to bring people indoors; and we can enact major lifestyle changes, including moving out of the affected areas, temporarily or permanently.

All these suggestions devour more energy and other resources, contributing further to climate change. And they are always easier for those with wealth already, providing more power (electrically and politically) to those who already have plenty of it. As usual, the people who most need assistance to deal with hazards—in this case influenced by the changing climate—are those least likely to obtain the support they require. People will suffer because we are not willing to provide what is needed for them; they suffer because of their high vulnerability.

What is the disaster here? The disaster is our unwillingness to use our abilities, power, and resources to help people deal with climate change. This is not to deny that some of the extreme weather arising from climate change could be debilitating and lethal, with few options available for avoiding it. The disaster nonetheless remains the vulnerability. People are placed in situations where they have no option but death in the face of the impacts of climate change.

It might be the same with the effect of anthropogenic climate change (that caused by humans) on sea levels, which are expected to rise, driven by three main factors. First, ice, snow, and permafrost are melting, with meltwater expected to raise sea levels by several centimetres during the twenty-first century. The second factor relates to the density of water. Some of the heat trapped by greenhouse gases in the atmosphere is absorbed by the oceans. Above 4°C, the density of water decreases, expanding the water's volume as it warms, which we see as a rise in sea level. By the end of the century, this heat-related expansion is expected to add more than a metre to sea levels.

The third factor has large uncertainties. Greenland and Antarctica are weighed down by enormous ice sheets. Climate change could trigger—might already be triggering—these giant slabs to slip slowly into the sea and melt, a process termed 'ice sheet collapse'. In worst-case scenarios, over several centuries, seas would rise higher than a twenty-storey building, drowning around a dozen island countries and many coastal cities. Much of Amsterdam, Manila, and New York City could be transformed into territory for snorkelling or scuba diving. Even if ice sheet collapse is more limited, we could still be looking at mass migration inland and the need to reconfigure world ports. We do not at present really know the likelihood of these catastrophes, but the consequences are monumental.

Could we counter climate change as an influencer of hazards? Yes, by stopping the major human influences on climate, through two broad categories of actions: reducing greenhouse gas emissions and increasing their uptake. A major way of reducing greenhouse gas emissions is by reducing our energy consumption, especially our use of fossil fuels. Healthier diets, more plant-based and focused on local resources, will also significantly cut greenhouse gas emissions. The uptake of greenhouse gases can be increased by measures that include stopping forest destruction and restoring forest ecosystems which have already been destroyed.

Interestingly, all these actions should be implemented anyway for reasons other than climate change. Reducing consumption makes sense because resources including fossil fuels are finite, so we will need to wean ourselves off them anyway. Healthier diets reduce healthcare costs and improve our quality of life. As for forests, in addition to absorbing greenhouse gases, they help to keep our air and water clean while providing long-term opportunities for people to generate livelihoods ranging from food to local tourism without destroying the ecosystem. Climate change adds a strong impetus towards these actions, yet so many other benefits accrue.

Adopting all these actions promptly does not mean that anthropogenic climate change stops immediately. We have already affected the atmosphere so much that even comprehensive and swift action will require decades before human influence on climate is far less discernible. Vulnerabilities must continue to be addressed—not just for climate change, but for all hazards and hazard influencers.

Unstoppable Hazards?

Could we really address all vulnerabilities to all hazards all the time? There are some massive, planet-wide perils for which the answer must be no, though even for hazards at this scale there are in many cases actions which we might take to reduce the extent of the disaster.

Impact by a large object from space, such as a comet or asteroid, could endanger much life on earth and might seem impossible to avoid. But we have options and opportunities to monitor and respond to imminent threats.[5] For space objects, we can actually do much more to tackle the hazard than to reduce our vulnerability, and baseline monitoring for potential threats is already happening. The main difficulty is convincing governments to join forces to deploy and maintain systems that would give enough lead time to act on an object coming from any direction at any speed. Luckily, trade-offs work in our favour.

An object's destructive potential comes largely from the product of its mass and speed. The higher the mass, the more damaging its impact will be. But large objects, and those moving at speed, are easier to spot, while the gravitational disturbance caused by small, dense objects would be detectable. Even a slow-moving, tiny, dense object slipping by all the planets with us in its sights would eventually perturb the earth and the moon.

We should be able to note its presence—if we are watching carefully, continually, and properly, showing the importance of investing in a complete and continuous space object monitoring and response system. Without this system, damage from an object smashing into the earth could be enormous. Even if it doesn't reach the ground, skipping through or burning up in the atmosphere on its way down, it produces a huge explosion in the air, raking the surface with its heat and pressure wave.

On the morning of 30th June 1908, Siberia's Tunguska region was shattered by a massive explosion, far greater than any known human power could have produced at the time. Trees were blasted as far away as 45 kilometres and a man 64 kilometres away was thrown from his chair. Seismometers in the UK registered the event. No impact crater could be found, because the object had seemingly detonated and disintegrated high in the atmosphere, delivering colossal heat and pressure to the surface. Vegetation at the site was completely charred. According to local folklore, some people were incinerated, but since no names were confirmed, fatalities remain a mere presumption.

Then, the morning of 15th February 2013 brought what might be the first verifiable evidence of human casualties from a space object. Brightening the Russian dawn, a meteor streaked through the air and burned up over the Chelyabinsk region, videoed by dashcams, phones, and security cameras. Within seconds, the shockwave shattered thousands of panes of glass, injuring over 1,200 people, and set off car alarms. Chunks of rock reached the ground, but no one reported any of them falling in populated areas or anyone hit by fragments.

A space object that reaches the earth's surface is called a meteorite. A strike on land emulates a missile and rocks easily penetrate vehicles and buildings. In December 1992, a 6.5 kilogram meteorite went through the roof and two floors of the house of a Mr Matsumoto of Mihonoseki, Japan.[6] He apparently thought at first that a thunderstorm had damaged his roof. Many media and scientific reports over the centuries ascribe deaths and injuries to being hit by a meteorites, but none are confirmed. The island of Saaremaa, Estonia, bears a crater lake from the Kaali meteorite, which fell several thousand years ago. Any people nearby would certainly have been killed, but the date of the meteorite strike and the date of human habitation of the island are both disputed, so we cannot be sure of any casualties from Kaali.

Objects larger than those hitting Tunguska, Chelyabinsk, Mihonoseki, and Saaremaa would be worse than our most destructive weapons and have been proposed as the cause of some of the mass extinctions throughout earth's history. Remnants of craters over a hundred kilometres across dot the planet from the past billion years. More have vanished under the constant tectonic recycling of rock. Impacts of such size would toss gargantuan quantities of soil, rock, and dust into the air, enveloping the planet and cutting off sunlight. World temperatures could plummet for years, wrecking agriculture and ecosystems.

A meteorite strike on water would vaporize the liquid below it and generate tsunamis, laying waste to coastlines and flattening cities far inland. A near-shore impact would leave victims just enough time to be numbed by the pressure wave and perhaps to register the water towering over them. A deep-sea landing might give hours of warning for evacuating millions. No human structure could withstand the brunt of the waves.

Rather than counting up such massive losses, surviving the aftermath, and reconstructing countries, warding off a strike is cheaper and easier. Unlike Hollywood movie scenarios in which humanity is saved from objects from outer space with nuclear bombs and individual, heroic self-sacrifice, reality is less histrionic. An object's path can be deflected to ensure that it misses the earth. The farther the object is from earth when the deflection is made, the less the deflection needed. At the right distance, the difference between a planetary catastrophe and a near miss is a fraction of a degree, which is why early detection and response is paramount. All plans for deflection demand tremendous monetary and energy costs, but remain far cheaper than the monstrous expense and fatalities of an impact.

We can, then, deal with space objects if we choose to do so. Any disasters resulting from them will be a consequence of our inaction; not a failure to eliminate vulnerability, but a failure to deal with the hazard. At least, most of the time. Tiny, high-mass, snail's-pace objects might exist, ramming us without warning to result in a scarred planet. Nature could produce planet-wrecking astronomical incidents which would be true 'natural disasters'.

It is not just nature's unknowns which might create rare 'natural disasters' because both hazards and vulnerabilities to them are hard to address. Known threats in astronomy include supernovae (the spectacular death throes of certain types of stars) and sudden radiation flares from others such as neutron stars. Our planet encounters serious trouble from such hazards perhaps every several hundred million years on average. Averages do not help us if it happens tomorrow. Nonetheless, the chance of a radiation flare or supernova within killing distance of earth is almost nil in the near future. We are getting to know the stellar environment around us in detail, without seeing any obvious, imminent dangers—at the moment, although if I am wrong, then perhaps no one will be left to tell me. Closer to home, the effects of our own sun's activity, described as 'space weather', is a hazard that we do know about and can plan for. The sun's atmosphere spews out a steady stream of charged particles which is called the solar wind. The solar wind strikes the earth's magnetic field and is diverted around the planet, with the particles being blocked except where they spiral in around the north and south magnetic poles, producing the dancing greens and reds of the auroras, the Northern and Southern Lights. As the strength of the solar wind varies, more or fewer particles arrive at the earth's surface. Intense activity on the sun can produce a geomagnetic or solar storm.

During a solar storm, the earth's magnetic field and ionosphere can be severely disturbed, disrupting electronics, frying power supplies, and stymieing communications. On 24th March 1940, space weather interfered with power supplies and communications in southern Ontario and the north-eastern USA. In October–November 2003, a large series of mass ejections from the sun led to the loss of one satellite and the shutdown of others, as well as interruptions of electricity grids and communication systems. Some polar airline flights were rerouted and warnings were issued to airline passengers and crews regarding radiation dosages.

These responses demonstrate how to reduce our vulnerabilities to space weather. Depending on the exact type of emissions from the sun, we could have hours to days of warning that severe space weather is likely and then minutes to hours to alert us that it is definitely happening. Solar storms might last for hours at most, although periods of high activity can continue for days and there are decadal highs and lows. Preparations must be completed long before a warning is issued, by adding shielding to electronics and being ready to shut down and restart equipment and systems;

in other words, designing the possibility of space weather into the systems, whether satellites or water treatment plants. People must also be ready for a power outage, by having their own emergency kit available beside stocks of non-perishable food, drinking water, medicines, and essential hygiene products.

The largest recorded space weather event from the sun occurred long before we depended on electricity, on 1st to 2nd September 1859. Named after the amateur astronomer Richard Carrington who made the connection between observations of activity on the sun and impacts on the earth, the Carrington Event severed telegraph communications in North America and Europe. An eruption of particles from the sun on 23rd July 2012 neared the level of 1859, but the main stream crossed earth's orbit at the point the planet had been just one week before. A mere seven days separated humanity from trillions of dollars of immediate damage to information and communication systems which might still not have been fully operational today.

While the earth's magnetic field aids in sheltering us from the solar wind and other radiation blistering through space, it too can form a hazard. For over three billion years, this field has been driven by molten rock circulating deep within the earth, with its intensity and direction constantly shifting, according to the vagaries of the core's behaviour. The field reverses polarity completely every million years or so, with the reversal taking centuries or millennia to complete. The most recent flip was just over three-quarters of a million years ago, with smaller shifts occurring since then, most recently 41,000 years ago. These changes long pre-date modern reliance on satellites, wireless communications, and electronic devices, leaving us vulnerable to the planet's magnetic field changes, as with space weather.

How much should our concern be directed at these hazards from space and from the earth's interior, and how much is our reliance on vulnerable technologies to blame? We have the technology to shield our power grids, satellites, bank machines, and phones from these environmental phenomena. Many of them have in-built safe modes which are activated when threats are detected. Providing full protection for all our earth-bound, atmospheric, and space equipment would not be easy because of the cost and time involved. It might not even be appropriate or worthwhile. When we design a satellite which uses the earth's magnetic field as an efficient mechanism to maintain its orbit, vulnerabilities to a shift in the magnetic field are in-built.

Many people today live without connections to these technologies, although suggesting that the rest of the world reduces reliance on them is not likely to be met with overwhelming support. Imagine being without social media for a day or not being able use a debit card at will! For some, this might feel like liberation—until they are affected by the essential services relying on electronic devices and communications including healthcare, transport, energy, food, and water. Even those without the latest navigational devices could end up lost, as old-fashioned hand-held compasses would be rendered as useless as GPS. The hazards in these cases are unstoppable with the line becoming blurred between our ability to address planet-wide vulnerabilities and our preference to live with them.

Another potentially unstoppable hazard that tests our capacity for curtailing vulnerabilities across planet earth is large-scale volcanism. These events sometimes involve flood basalts that ooze lava dozens of metres deep from fissures tens of kilometres long. The Siberian Traps flood basalt erupted approximately

250 million years ago, covering millions of square kilometres, with some estimates calculating its area to be larger than India today. The Siberian Traps and some other flood basalt eruptions occurred at about the same time as mass extinctions across the planet, and may have been a cause, especially given the major impacts that such vast and extended eruptions must have had on global climate.

Individual supervolcanic eruptions can also knock the global climate into a different trajectory. Ash injected high into the atmosphere spreads out, encircles the globe, and for years reduces sunlight reaching the earth's surface. About 74,000 years ago, the skies above what is now Lake Toba in Sumatra, Indonesia darkened, thunderous roars rolled over land and sea, and pressure waves dissipated around the world. The biggest known volcanic eruption in the past 2.5 million years burst upon our ancestors. Ash from the Toba super-eruption has been identified 7,000 kilometres away, with the associated gas signal appearing in air bubbles trapped in ice cores from Greenland. A near-extinction of the human species around at the time was initially attributed to global cooling caused by the eruption, but subsequent studies did not find evidence that the super-eruption either produced a prolonged volcanic winter or posed a major threat to humanity. A bigger supervolcanic eruption could bring both, meaning that we are fortunate that one has not happened recently.

Ice is another global hazard. Many ice ages have come and gone, driven by variations of the earth moving in its orbit. Kilometre-high glaciers bore down on, and later melted from, places that now house megacities. Any ice age today would force millions to flee as homes disappeared under the white masses. This amount of ice had to come from somewhere: the oceans are much higher today than when glaciers scoured the planet. The latest ice age peaked about 20,000 years ago, at which point sea levels were over a hundred metres lower than now. Some of today's coastlines sat hundreds of kilometres inland.

From exploding stars and meteorite strikes to ice ages and volcanic madness, planet-wide vulnerability is ever-present and cannot be eliminated. The human species could be threatened by large-scale environmental events and processes. Some may seem fanciful, but they all provide realistic scenarios and have happened in the past, though long before humans appeared. Little scope exists to reduce vulnerability extensively to many of these hazards, even with ample warning, and to tackle such hazards. These might truly be called natural disasters.

Why Choose to Create Disasters?

Apart from rare planet-wide hazards, nature produces myriad hazardous and potentially hazardous events and processes. If we are not aware of them, they can appear to be unusual, extreme, unpredictable, and impossible to tackle. But even when we are aware of them, decisions may be made that are tantamount to choosing to create a disaster.

Ask many people about earthquakes in the USA and they react with 'California!' The San Francisco earthquake and fire of April 1906 was immortalized in prose by survivor Enrico Caruso, one of the world's top opera tenors of the era. Los Angeles is typically listed as a city awaiting a big shaking. With San Francisco and Los Angeles having made it through major damage from, respectively, tremors in 1989 (the Loma Prieta earthquake) and 1994 (the Northridge earthquake), it is unsurprising that California is, in the public consciousness, the Earthquake State.

The high hazard zone extends northwards. Oregon, Washington, British Columbia, and Alaska have all gone through massive shakings in the past and await more in the future—as have the countries southwards starting at Mexico. Portland, Oregon, and Vancouver, British Columbia have yet to be tested recently, whereas Seattle, Washington was rocked moderately in 2001. On 28th February at 10.54 a.m. locally, the Nisqually earthquake, sixty kilometres south-west of Seattle, entered the books at moment magnitude 6.8 with a depth of fifty-two kilometres. One heart attack death is attributed to this shaking, although several other cardiac-related deaths in the forty-eight hours afterwards might be linked to it, and about 400 people sought medical assistance for injuries.

This is clearly a success story, with no fatalities described from collapsing structures. How did it happen? Since the area's previous major earthquake, in 1949, building codes and construction practices had implemented seismic safety. Bridges, especially, had been built to new earthquake safety levels for twenty years before 2001, alongside a retrofit and upgrade programme for existing bridges. The few post-earthquake bridge closures were mainly of older structures, many of which had already been designated for improvements. The state legislature, two large office buildings, and the air traffic control tower received substantial damage, yet did not kill anyone. One other factor underlying the low death rate was that an area of Seattle which suffered significant earthquake damage had been closed off to the public due to riots the previous night. Older buildings in the Pioneer Square and SoDo Historic Districts rained down bricks, with the potential to kill or injure, onto empty streets.

All the same, Seattle has developed a culture of readiness. In 1971, the city began to run free three-hour CPR training sessions for residents. The impact of first aid training is evident from the lives saved by early, safe, bystander interventions. Such training may also instil an attitude of preparedness and prevention before a crisis, though links between basic training and long-term local spirit rely principally on anecdotes and are hard to confirm.

More tangibly, Seattle had embraced Project Impact, started in 1997 by the US federal government to bring together residents and local businesses to prepare for hazards. From structural upgrades for buildings to information campaigns, Project Impact fashioned and financed locally led vulnerability reduction ventures. Seattle adopted it to entice homeowners into earthquake retrofits and to modify public infrastructure to withstand shaking. One school stated that during the 2001 earthquake, a large water tank did not smash through a roof into an occupied classroom because they had made it safe thanks to Project Impact.

Just a few hours before Seattle heaped praise on Project Impact given the lives it had just saved during the tremor, US President George W. Bush terminated the project in his budget plans, a decision eventually enacted. Why decide to create disasters by ending an evidently successful vulnerability reduction initiative? Most commentary suggested that a Republican president could not accept positive results from the previous Democratic administration. Despite being cost-effective, encouraging private–public partnerships, and devolving responsibility from the national government—principles claimed by many Republicans as the ethos of less government 'interference'—it seems that not even demonstrably preventing human mortality was enough to save Project Impact from a politically induced death knell.

Seattle's success of February 2001 does not ensure future safety. This earthquake's magnitude was over a hundred times less than the magnitude 9+ upheavals that have convulsed Washington State in the past, not to mention tsunamis. From south of Los Angeles to north of Vancouver, the work never ends to reduce vulnerability for forestalling disasters.

The discussion above has concentrated on the west coast of North America. Fly five or six hours across the continent to the east coast and hopefully the worst we need to contend with is a Category 5 hurricane. Except, as it turns out, earthquakes rumble along the USA's Atlantic coast too, albeit not generally as big as those affecting Pacific cities. Conversely, California has much more experience and is far better prepared than the other side of the country. Vulnerabilities are much higher for a smaller hazard on the east coast compared to the west.

Soon after the east coast's lunchtime on 23rd August 2011, a moderate, shallow earthquake of moment magnitude 5.8 and depth of six kilometres originating in Virginia was felt by about one-third of the US population. The North Anna Nuclear Generating Station in Virginia shut down, masonry fell off houses, skyscrapers swayed in New York, and viewers watched news anchors swear or freeze in their seats. When completed in 1884, the Washington Monument was the world's tallest building. On the day of the 2011 earthquake, security cameras atop the obelisk captured tourists inside dodging falling stones and mortar as they ran for the stairs. It required several years and a private philanthropic donation to repair the earthquake cracks and other damage. No one seems to have died as a result of the tremor, but it revealed the abundant earthquake vulnerabilities pervading this part of the country.

In Boston, emergency services were called to a purported building collapse which turned out to be a false report, but may presage the city's future prospects in tremors. Two damaging earthquakes struck Boston in the seventeenth century and two more in the eighteenth century. Since then, notwithstanding August 2011, seismic activity in the region has been comparatively low, while seismic vulnerability in the area has ratcheted up.

On 18th November 1755, Massachusetts experienced the most intense earthquake recorded since Europeans are known to have reached its shores. A woodcut from the time illustrating the event shows buildings a maximum of four storeys high, with church steeples collapsing. Most damage in the 1755 earthquake seems to have been to structures built on land formed by dumping earth into the harbour. Such infill is, as mentioned earlier for Mumbai, notoriously unstable during seismic activity, because seismic waves cause the ground to liquefy.

Boston today has substantially expanded the infill areas. Inland, car-friendly suburban sprawl consists of detached houses and low-rise apartment blocks. By contrast, the coast has seen billion-dollar developments of high-rise condominiums, with prices starting at just below seven figures in US dollars, in what was coined the 'Manhattanization' of the city. The skyline sports glass-fronted tower blocks, cloaking a greenway through the city centre sitting atop a major motorway tunnel which opened in 2006 following substantial cost and time overruns and technical faults. This 'Big Dig' and the new buildings above it have yet to be tested by a significant earthquake.

Depending on how the infrastructure and population respond, a repeat of the 1755 earthquake in Boston could be an inconvenience, a devastating blow, or anything in between. While much remains to be understood about earthquakes in and around Massachusetts, history combined with contemporary investigations suggest hazard characteristics which are likely to shake the city.

The consequent disaster, or lack thereof, is unknown because vulnerability is the wild card. Even if parts of the Big Dig collapse, the outcome would be very different if it occurs during rush hour with thousands of people in their vehicles rather than in the middle of the night, when most people are at home in bed. The location and amount of liquefaction interacts with the construction of the buildings to make the difference between some tilted structures from which people can escape and a recovery operation taking weeks to pull out hundreds of bodies. Would a single collapse in the downtown core set off a chain reaction bringing down more structures? How is people's behaviour changing from the city's earthquake drills and information offensive?

Boston's earthquake is certain to happen at some point. The city's earthquake vulnerability is present and visible, although the potential range of disaster outcomes is huge and its bounds are hard to pinpoint. The Boston earthquake disaster resides firmly in the realm of the unknown, depending much more on what the vulnerability turns out to be rather than on the specific earthquake parameters.

The same story is told farther south along the eastern seaboard. In August 2011 as in November 1755, Charleston, South Carolina felt the shaking. Today, colonial and modern architecture mingle throughout the starting place of the US Civil War, and its population, estimated at just shy of 150,000, is scattered across its various peninsulas and islands. The local government offers plenty of advice for dealing with hazards, naming fires, floods, storms including hurricanes, and earthquakes. Recording ten to twenty earthquakes a year across its state, Charleston, like Boston, is actively involved in earthquake-related drills, information, awareness, and preparedness. How truly ready is the population?

Charleston's recorded list of earthquakes dates back to 1698, with one of the largest recorded tremors along the US east coast occurring on 31st August 1886. More than five dozen people died out of Charleston's population of 50,000 at the time, with reports of damage as far away as New York City. The damage to each of Charleston's 7,000 buildings was published, with brick faring far worse than timber, and twenty-three buildings were noted as requiring demolition.[7]

In continuing parallels with Boston, sections of Charleston are built on reclaimed land prone to liquefaction in an earthquake. Both cities are aware of the hazards and are making efforts to reduce vulnerabilities. As in every other such city, there is still much to do. And as in other such cases, how effective are the efforts in Boston and Charleston to accept and address the fundaments? Would it even be feasible to stop building on areas prone to liquefaction and to fully shore up the structures which are there already? How could threats from tsunamis (which might be set off by earthquake-triggered landslides) be dealt with for those who live or work in potential inundation zones?

Delving deeper, how possible is it to identify and confront the baseline creators of vulnerability? Where would we begin in tackling gaps between rich and poor, inadequate education, and discrimination against and marginalization of many groups?

For many, the relatively rare possibilities of earthquakes and tsunamis seem far removed from the overwhelming concerns, dangers, and risks of coping with day-to-day life. Add in planning regulations, building codes, and construction techniques which could always be improved, and we can see why disaster prevention is so challenging.

In planning discussions, we often hear that the work is expensive, and the best outcome is being achieved despite constrained resources, given all the other pulls on budgets and attention. The claims are that vulnerability reduction is not being deliberately evaded and that problems are not being specifically created for those who cannot help themselves; rather, it is a matter of cost. But this excuse does not withstand careful scrutiny. Detailed examination invariably demonstrates that the money exists to tackle disasters if we choose to apportion it to this task. Instead, we choose to prioritize other activities, producing and maintaining much more disaster vulnerability than we reduce.

Worldwide expenditure on the military is declared to be around US$1.5 trillion per year—around ten times higher than international development aid. About as much money might be spent on undeclared military spending or on military-related expenditure from other budgets. Illicit financial flows—referring to money or capital earned, used, or transferred illegally—from less affluent countries to more affluent countries total over six times as much as international development aid. If rich people in less affluent countries would invest in their own societies, then we could eliminate international development aid and the rich people would still have plenty of cash left over. What possibilities exist for the banks and tax havens, typically in more affluent countries, to stop these fund transfers and to encourage investment in people and society?

In the meantime, many polluting industries continue to be heavily subsidized by governments using taxpayers' funds. Direct fossil fuel subsidies are calculated at approximately US$600 billion each year. When health and environmental impacts are considered, this value is multiplied sevenfold. Tax avoidance and skewed tax regimes favouring the already rich are being increasingly implicated in resource misdistribution.

All these expenditures and allocations are justified by those who make them. Without judging whether or not their arguments are convincing, the amount of money involved demonstrates that the resources are available for disaster prevention activities, including some larger-scale technical initiatives, such as a complete response system for space objects. More to the point, it demonstrates that reducing vulnerability through social change could be widely achieved. Instead, active choices allocate the money available to activities creating vulnerability.

The minority with wealth and fiscal and political power decide. Some are elected to their positions; some acquire their power through skill, talent, and/or hard work; and others inherit their positions or use deceit to get there. Some retain a modicum of accountability through shareholders, genuine elections, or consumer choices. Many more earmark resources by themselves for themselves, actively avoiding accountability including through tax evasion.

These choices not only lead to disasters, but also cost money. If the elites really wished to save as much money as possible, then they would advocate for eliminating fossil fuel subsidies and tackling climate change. Providing indoor temperature control for everyone who will swelter in intense heatwaves as the earth's climate patterns shift is extravagantly pricey. It requires altering infrastructure around the world, from Jakarta to Baltimore and from London to Rio de Janeiro—as well as in the Sahel. Even in cities with widespread indoor temperature control mechanisms already, or prospects for them, such as Baltimore, poor access could inhibit implementation. Not everyone can afford the electricity bill; the maintenance, repair, and upgrade costs of the equipment; or retrofits for existing buildings where needed. Many could not afford to stay indoors during heatwaves, because they need to go outside for food and water, and to work. As usual, the worst impacts of disasters, and the fewest options for improving their situation, fall mostly on those with the fewest resources and the least power.

We can now answer the question, 'why do some people let disasters happen by creating vulnerability?' Because most of the vulnerability they create is for others. Ultimately, a minority creates vulnerability, and hence disasters, for the majority, because they do not care or choose not to be aware that they are doing so. And, ironically, in creating vulnerability for others, they also create some for themselves, since vulnerability is so interconnected.

We could allocate resources better. This does not mean throwing all our money at activities. It refers to investing strategically to reduce vulnerability at its baseline causes while ensuring monitoring of and accountability for the resources which are invested. Part of the task is allocating power to be more equitable and fairer, again with balances for monitoring and accountability. Nor does it mean that we should readily condemn or condone specific systems of government or governance; most have benefits and flaws.

The point is to admit those benefits and flaws, be honest about the vulnerability and disaster consequences of resource and power decisions, and map out who decides and who is affected, particularly noting how small groups reap the rewards of their decisions so that others without as many options must bear most costs.

We know what to do to stop ourselves causing disasters. Not everyone wishes to make these choices, or has the power to do so. Vulnerability cannot be eliminated unless those who can, choose to act.

CHAPTER 6

MAKING THE CHANGE

Success in Toronto

Toronto sprawls along the north-west shore of Lake Ontario, a typical North American megacity with a downtown core, and other focal centres, surrounded by straggling suburbia. As Canada's most populous city, the metropolis today buzzes with languages, food, and music from around the world. People are drawn there to enjoy the economic opportunities, comparative safety, and high quality of life it affords. This quality of life includes the accommodation of nature within the city. Meandering rivers carve cliffs and ravines through residential and commercial districts, eventually emptying into the lake. These waterways flow as placid streams during regular or dry weather, morphing into whitewater death traps under intense rainfall.

Four thousand kilometres away from Toronto, on 5th October 1954, a tropical storm coalesced south of Barbados. Less than a day later, it graduated to hurricane status, was dubbed Hazel, and moved west through the south Caribbean Sea. Why would anyone in Ontario, thirty degrees of latitude north of this track, be worried? On 9th to 10th October, the storm turned sharply right and headed north. Eerily prescient of hurricane Matthew's trail in 2016 more than two generations later, hurricane Hazel made a beeline for Haiti in the north Caribbean, yielding a death toll similar to that of Matthew. After Haiti, Hazel advanced north, traversing North Carolina and still going strong.

On 15th October 1954, hurricane Hazel crossed Lake Ontario and slashed through Toronto, killing eighty-one people. Rivers transformed into raging torrents, sweeping away houses and their occupants. Over a third of the deaths occurred on a single street: Raymore Drive. This unassuming road was in a prime location along the Humber River, a peaceful riverside setting—until the river rose and quickened with Hazel's rains. Homes and families vanished into the frothing waters.

Houses along a third of a kilometre disappeared overnight. One resident returning from a party elsewhere led families to safety, to discover the next morning that his relatives had perished. Firefighters were long haunted by the screams, slowly fading or suddenly halting, of those who could not be reached through the howling wind and rushing water. Spotlights shining on some of the trapped individuals illuminated only the impossibility of reaching them, because of a lack of training and equipment.

In the aftermath of the funerals and tales of horror, Toronto and its surrounding region reorganized the planning for and management of river floods. A new layer of municipal government called 'Metropolitan Toronto', formed in 1954 and dissolved in 1998, took responsibility for urban planning. Their tasks covered public transport, major roads, water, sewage, police, and coordination for dealing with floods, with firefighters remaining governed by the lower level of municipal government.

Toronto converted residential roads ravaged by hurricane Hazel into parks and recreational areas by reinstating them as floodplains. Raymore Drive, the part of it that had not been annihilated, still parallels the Humber River, but on a bluff several metres above the water. The road slopes down towards the river, ending where the carnage of 1954 began. Rather than houses at the bottom of the slope, Raymore Drive gives way to greenery, a playground, a walking and cycling path, and a poignant memorial to those who lost their lives.

Toronto today is renowned for its lengthy greenways (which just happen to be floodplains) used for walking, cycling, picnicking, enjoying nature, and environmental education. The Humber River Trail, on Toronto's west side, has perfect spots for photographing salmon jumping the small weirs as they swim upstream on their autumn run to breed. The Don River Trail, in the east, places bridges conveniently for salmon spotting (and playing Pooh sticks) in the river underneath. Anyone proposing riverside development would be ostracized, not because it would place buildings in areas vulnerable to floods, but because it would mean destroying much-loved nature and recreation areas.

Subsequent hurricanes and storms affecting Toronto, like hurricane Isabel in 2003, burst riverbanks with the water felling fences and trees as mud and debris slithered across the paths. But these were easily fixed or cleaned up. No widespread flood damage occurred, because of the absence of property in the floodplains. Toronto had reduced its vulnerability to hurricanes, saving lives. A storm became an inconvenience, not a disaster, through human choices.

Toronto's flood story does not end here; in fact, it never ends. The city has changed substantially since Hazel's rains abated, particularly from continuing urban expansion which paves over farms and other green areas, augmenting runoff into the ravines. A huge motorway, Highway 401, was initially constructed as a city bypass to the north. Completed in 1956, it now cuts east–west directly through residential and commercial parts of Toronto, with multi-lane bridges over the Humber and Don rivers. An urban conurbation termed the Greater Toronto Area houses over six million people, more than quadruple the area's population which survived hurricane Hazel.

The 401 serves this urban monolith, as do other major motorways including Highway 404 which, as it crosses the 401 going south, becomes the Don Valley Parkway or DVP. Its name comes from the Don River, which the motorway parallels, snaking its way south to the lakeshore by running near or alongside the Don's ravine, the Don Valley. Also matching the twists and turns of road and river, train tracks take commuters from the northern edge of the Greater Toronto Area to Union Station, Toronto's central train station, in the middle of downtown near the lake. It is important that the DVP and the railway run 'near or alongside' the Don Valley, not in it. Both the road, completed in 1966, and the train tracks, first opened in the nineteenth century but not fully finished and used until 1978, generally sit above the level of the river and mainly outside the floodplain.

The situation changes at the foot, or south end, of the DVP, where both road and railway border the Don River along a flat stretch of around two kilometres. This is floodplain. The hazard was known at the time of construction, and the planners accepted that the road and tracks would flood on occasion, blocking the routes for a few hours as the water drained into the lake. The same holds farther north, near a place beside the river known as the Brickworks, where the tracks dip to about the same elevation.

The known flood hazard and several previous floods along the tracks, as well as a train trapped in floodwaters in 1981, did not stop a double-decker passenger train from stalling in floodwater on 8th July 2013, as it was trying to reverse away from rails that were becoming rapidly inundated. Rescue boats were deployed, to pull those with medical conditions through the windows. One passenger who was clinging to trees after having tried to swim away was also carried to safety. Over six hours later, the floodwaters had subsided enough to allow the remaining hundreds of passengers to take a muddy walk to safety. Had the water been deeper, faster, or stood for longer, the ending might not have been so happy. Despite the history of flooding, only after this incident were flood alarms installed along the tracks—now, dispatchers can stop or reroute trains when water touches the railway.

For the DVP, in the mid-2000s the City of Toronto expressed interest in using accurate river flow information from a regional agency to anticipate flooding and to support timely response. The idea focused on installing gates on ramps accessing the highway. These gates could be closed electronically from central control or manually, to stop drivers from entering the floodable portions. But in the end, discussions could not progress this initiative.

Similar approaches were re-explored after the 2013 train incident, and debates about dealing with the flooding of the DVP and the train tracks continue. Raising them would be expensive and disruptive, putting major commuter routes out of action for weeks, if not months, while rerouting would introduce numerous complications. With continuing urban development and expansion, flooding and closures will not go away.

Toronto's flooding concerns are not confined to the Humber and Don. Flash flooding on 12th May 2000, 19th August 2005, 25th June 2014, and 17th August 2018 among many other dates, submerged basements and underpasses around the city. Sometimes, drivers had to be rescued after they drove through the water and became trapped. Although brief, the floods have in some cases caused tens of thousands of dollars of damage per house.

Major infrastructure changes such as removing basements and underpasses are not simple and we might not want to remove them. Basements serve as guest quarters or rental spaces for extra income and underpasses support efficient travel for vehicles, bicycles, and pedestrians. Still, as we have seen, education about common sense behaviour for hazards, such as not entering floodwater—no matter what vehicle we are driving and not even as a pedestrian or cyclist—can go a long way towards diverting people from danger.

The summer of 2017 reintroduced the city to yet another flood threat, this time just offshore, across the small archipelago called Toronto Islands. The three main islands, just a fifteen-minute ferry ride from Toronto Harbour, are subdivided into a multitude of islets with a maze of walking paths surrounding picnic areas, yacht berths, sports fields, and a disc golf course. At the western end, Hanlan's Point is dominated by the regional Billy Bishop Airport, reached by a ninety-second ferry ride or a 260-metre pedestrian tunnel. Centre Island, in the middle of the archipelago, has a summer fun fair which is a regular day out for families and children.

On the eastern side, Ward's Island, an islander community of over 600 people in 262 houses, started out as a fishing village in the 1830s. Today, visitors can take narrow, flower-lined paths amid the picturesque cottages from the beach facing the lake to an isolated picnic table with a panorama of Toronto's skyline featuring the iconic CN Tower—once the world's tallest building and now claiming to house the highest wine cellar in the world—soaring above the glass-fronted office blocks of Canada's financial centre. Peaceful solitude in the middle of a world city, until the calm is blasted away by the roar of an airplane revving for take-off from Billy Bishop.

As April turned to May in 2017, the level of Lake Ontario rose, driven by above-average snowmelt earlier in the year and larger outflows from upstream Lake Erie, combined with substantial rainfall across and around Lake Ontario. Meanwhile, management of the lake's levels via a downstream dam ended up becoming complicated by patterns of ice formation during the early spring period. In late May, Lake Ontario's level was the highest that it had been in a hundred years, and development across Toronto Islands had expanded over this time.

Even before the water had reached this point, a flood disaster was in progress. On 4 May, the islands were closed to visitors and evacuation plans for residents were in place, awaiting the order. Mid-May witnessed beach closures and cancellations of permits for recreational activities. Just over a week later, half the land area of Toronto Islands was reported as being underwater with sandbags and pumps protecting the houses. Gardens transformed into bogs as the water table rose. It took until 31st July before tourists and mainlanders could once again hop on board the ferries, but visitor numbers never reached those of previous years.

That summer was a time of readiness and stress for those with homes and livelihoods across the archipelago, but it was not the first time that Lake Ontario had risen. Some living on Toronto

Islands in 2017 had made it through the floodwaters in 1973. Following the Second World War, lake levels almost as high as in 2017 occurred in 1951, 1952, 1974, 1976, and 1993—the conditions of 2017 hit after a generation-long hiatus.

Throughout the 2017 inundation, Billy Bishop Airport continued operations. It had closed during the July 2013 and August 2018 storms when an overwhelming amount of rainfall led to water pooling inside the terminal buildings. But because the airport elevated its land over the years, it managed to get through the summer of 2017 remaining relatively dry. A few centimetres of height made the difference between damaging flooding inter-rupting business and unabated operation. Many island residents and environmentalists wish that the airport would be disrupted regularly and, preferably, closed permanently to eliminate noise and air pollution. In any case, elevating construction would not be suitable everywhere, for instance in the case of beaches that slope gently to the waterline to support recreation. And altering the topography through local elevation must be done with care, as it inevitably impacts the surrounding environment, including how surface water and lake currents flow—and how other areas become flooded.

Major engineering works, such as walls and levees, could be constructed around the islands' homes, the fun fair, and other rec-reational areas. But unless designed carefully, integrating with the natural landscaping, they could substantially degrade the quality of life and recreation across Toronto Islands. And without due attention, they could interfere with the natural interactions and flows of water, land, and sand, causing more damage. They would never guarantee immunity from flooding anyway.

Neither do Toronto's ravines. How long might it be until storm water and erosion encroach onto houses backing onto these parks? Invasive species, air and water pollution, and the rapidly changing climate stress the ecosystems. Will the animals and plants respond with ecologies which tend to moderate intense rainfall and periods of precipitation deficit, or will new regimes augment runoff and scouring? What happens to floods, and to the city, if wildfire becomes common in Toronto's greenways as the air warms? What is the ravines' highest capacity for taking up rainfall, compared to the intensity of a superstorm which could envelop the city?

Toronto's flooding vulnerabilities and flood disasters have not faded away post-Hazel. They never have been, and never will be, static. Despite its city-within-a-park branding and despite the momentum from the lessons born from the hurricane catastrophe, minor and major flooding perpetually impacts the entire Toronto region, together with other hazards. Disasters are still possible and are still happening, with potentially more in the future due to ever-present and ever-rising vulnerabilities. By tackling the vulnerabilities, including drawing on the lessons of history such as from hurricane Hazel, we can choose to avoid these.

Success in Bangladesh

Southern Bangladesh is a world away from the middle-class brick houses of Raymore Drive. Thatch and corrugated metal roofs intermingle on coastal and river sands that shift with the tides and the storms. The people rely on agriculture and fishing. During the monsoon season, many roads become impassable mud and the area bears the brunt of cyclones as they sweep off the Bay of Bengal.

Cyclones are not uncommon in Bangladesh. Cyclone Sidr tore through in 2007, followed by cyclone Aila two years later, each causing severe damage and loss of life. Several other major cyclone disasters have occurred within living memory, in 1970, 1985, and 1991. These tragedies led to top-down work by the government to reduce cyclone impacts, including the construction of a series of coastal embankments interceding between land and sea. When cyclone warnings are issued, they reach people swiftly through various media, and the people understand what they mean and where they must evacuate to, mostly a series of robust shelters.

Not everything can be achieved at the national level. From 2013 to 2016, two Bangladeshi coastal villages, Nowapara and Pashurbunia, participated in a vulnerability reduction programme funded and backed by the British Red Cross and Swedish Red Cross and implemented by the Bangladeshi Red Crescent. The focus was not so much on rain, wind, and water, but rather on avoiding disaster impacts by integrating a variety of actions into day-to-day life. People were paid from the programme to construct better roads, to build latrines and water wells, and to explore ways of diversifying their assets and income sources. They received training for developing businesses and accessing markets. Some of the villagers now use their back gardens to grow vegetables to sell while others make clothes or rear cattle for meat and milk. Small shops have sprung up.

Dealing with disasters is also tackled directly. Fishers received safety equipment and residents learned first aid and search-and-rescue techniques. Identifying and treating diarrhoeal disease vastly improved, as did hygiene and sanitation. The poorest households—those without land or with marginalized members such as those with disabilities—were prioritized for receiving cash grants. Because of this programme, villagers sought ways to protect their lives, assets, and livelihoods during storms.

These actions apply beyond cyclones, increasing day-to-day community safety and adding income and options to the people's endeavours. They engage families who were not well connected beforehand to others around the villages. Fundamentally, the programme aspired to circumvent the need for spending money on post-disaster assistance and then calculated that for every taka (Bangladesh's currency) spent setting up the initiatives, almost five taka were soon paid back through increased earnings or reductions to other costs.[1] From a purely monetary perspective, this enterprise on avoiding disasters was financially successful—and a hazard had not even manifested!

But this creates a challenge. Without a hazard, can we really claim success and would we really know that vulnerability has been tackled and reduced? For this project, we need not speculate. Cyclone Ruanu made landfall on Bangladesh's south coast on 21st May 2016, just three weeks after this project officially ended. Those living in Nowapara and Pashurbunia received warnings, accepted them, and evacuated, with no casualties reported. They soon returned home to find that their new initiatives had withstood the cyclone's forces. Their new freshwater supply and latrines remained functional and they easily resumed most of their liveli-hoods. One embankment had broken, flooding some of their agri-cultural fields with salt water. Because the project had diversified livelihoods, many of the villagers were able to switch to fishing while they worked the land to help it recover from salt contamin-ation and return it to cultivation.

Nowapara and Pashurbunia had reduced their vulnerability to cyclones, saving lives. A storm became an inconvenience, not a disaster, due to human choices. How long might this success last? Could it be maintained until a local or national election shifts priorities and undermines the work? Or might the lessons remain with the people, to be handed on to the next generation and perpetuated as new leaders gain power? What about confounding factors, such as Bangladeshi government interventions to resettle to coastal areas Rohingya refugees who fled from violence and genocide in Myanmar in 2016–17? What knowledge do Rohingya have of hazards in the areas of Bangladesh where they might end up living?

An intergenerational continuation of the vulnerability reduction practices, along with new arrivals learning them, would demonstrate a remarkable success story. Livelihoods would further improve while many cyclones would traverse the area without disasters emerging, although as the decades unfold, so will social and environmental changes. Will the people of Nowapara and Pashurbunia remain content with fishing, agriculture, and small enterprises, or will the added opportunities increase the chances of youth drifting away for other careers in the cities or abroad?

Population movements and depletion of the best and the brightest might be a particular concern under presumed environmental changes. Coastal Bangladesh is said to be at the front line of climate change impacts, especially with respect to sea-level rise. The villages and the land around them are expected to be facing higher waves and storm surges as the next generation comes of age. Due to climate change, cyclones in the Bay of Bengal are projected to decline in numbers while increasing in intensity. How well will the embankments withstand the rising ocean's ferocious pounding?

In the absence of recurrent cyclones, will embankments and safety measures lapse, with the people creating their own vulnerabilities to the next, increasingly powerful storms?

Even more concerning is fresh water. Salt water infiltrating the wells could wreck the communities without a full water treatment system. As well, the embankments could be expanded and pumps could be installed to keep the land dry and salt-free. For how long would these technical fixes function, especially given the uncertainties in how climate change's impacts will manifest locally? How much energy, spare parts, and maintenance would be necessary? Could the villages afford to do what they wish?

Similar questions must be asked for other hazards facing Bangladesh. The country is prone to earthquakes, with little having been done to work on the population's or infrastructure's ability to deal with large, shallow tremors. How would the embankments fare under seismic shaking? Climate change is impacting Nowapara and Pashurbunia now and will not go away in the near future, but earthquakes and tsunamis could strike parts of the country at any time.

In addition to hazards from the ocean and underground, the villagers need to consider those from upstream. River characteristics vary according to activities all along the waterway's length, in particular runoff, dams, and water use, all of which affect the water volume, flow rate, and sedimentation. Shortly after Bangladesh's independence in 1971, India started operating the Farakka Barrage across the Ganges near the border between the two countries, diverting large flows towards Calcutta to increase river navigability there. Nearly two decades later, Bangladesh chimed in with the Teesta Barrage across the Teesta River in the north of the country, near the border with India.

These barrages substantially affect downstream salinity, sedimentation, and the availability of water in the dry season across Bangladesh. As climate change augments glacier runoff from the Himalayas over coming years and decades, the three great river systems of Bangladesh—the Ganges-Padma, the Teesta, and the Brahmaputra—will undergo further major changes. In the long term, what will happen to the Himalayan glaciers, to runoff from them, to rainfall, to dams, and to society?

The people of Nowapara and Pashurbunia have no control over these large-scale, long-term alterations, from India's actions along its rivers before they enter Bangladesh, to the impacts of climate change on those rivers' headwaters and tributaries, to large movements of refugees into or out of Bangladesh as well as internal migration around Bangladesh. Still, with external support and internal impetus, the people can address some aspects of their livelihoods and vulnerabilities, which is how the Red Cross/Red Crescent project began. No matter what the hazards or how they evolve, the people will have taken some major steps towards avoiding disaster for themselves. They can never be completely certain that they are disaster-free, but they will have tackled the true origin of disasters: the stage at which vulnerabilities are created.

The day may yet come when the encroaching sea cannot be stopped, or when external choices change the river water and sedimentation to the point that Nowapara and Pashurbunia cannot be sustained. Generational migration, starting livelihoods anew, and village reconfiguration could become essential. At this point, having built up capabilities and having reduced vulnerabilities over years, at least the people will be making their own decisions using their own resources from a position of strength of having learned and succeeded, rather than from weakness, destitution, ignorance, and hopelessness. No matter what the future might bring, the work done with and by the villagers of Nowapara and Pashurbunia will have helped them towards a brighter future with fewer disasters. It comes back to human choices that can reduce vulnerabilities, build self-reliance, and widen the resources and choices available to people directly impacted by hazards.

People Preventing Disasters

Human choices cause disasters, so human choices can prevent disasters. That is the main message of this book. We already have the knowledge, understanding, theory, and examples to show that we can triumph over ourselves. When will we have the wisdom to do what we know needs to be done everywhere, learning from those who have succeeded?

The year 2003 brought seismic events of striking contrast. Early in the morning of 26th September, a massive earthquake of moment magnitude 8.3 with a shallow epicentre just twenty-seven kilometres below ground struck Japan's northern island of Hokkaido. A massive aftershock followed just over an hour later. No one died in the shaking, despite hundreds of injuries. Reports suggested that one person was killed when they were hit by a car while sweeping up glass.

Three months later, across the Pacific during the late morning of 22nd December, central California experienced a moderate tremor of moment magnitude 6.5, very close to the surface, just eight kilometres deep. The death toll reached two when a tower collapsed as people were trying to leave the building, leading to a court judgement against the property owner for inadequate reinforcement and maintenance. As with Japan, there were also numerous injuries and plenty to clean up.

Just four days later, a similarly large earthquake, of moment magnitude 6.7 at a depth of ten kilometres, hit southern Iran just before dawn. Here, over 25,000 people died. Many suffocated as their mud brick houses turned to dust or they were crushed when heavy roofs came down on them. Had the earthquake struck just a few hours later, when people were in the streets travelling to work and school rather than being inside, perhaps the death toll would have been vastly different.

The difference between California and Japan on the one hand and Iran on the other, in 2003, was not so much the earthquakes, which are merely hazards, but the vulnerabilities, starkly illustrated again in Japan on 11th March 2011. Off the coast of Sendai in the north, the shaking registered a moment magnitude of 9.1 at twenty-nine kilometres depth, placing it within the top five most powerful earthquakes ever recorded. The ocean floor shifted by sixty metres in places. High-rises in Tokyo, three hundred kilometres away, swayed back and forth—which is exactly what they were designed to do. Oscillating in a planned fashion stops them from collapsing, burying thousands.

Back near the epicentre, buildings performed similarly well. They were designed to withstand earthquake shaking and they did, protecting their inhabitants and avoiding a major earthquake disaster. The response to the earthquake represents a major success story in dealing with disasters: a country knows the hazard, decides to prepare society for the hazard, is tested by an enormous instance of the hazard, and survives without substantial damage. Except that nature was not yet done for the day.

The reshaping of the seafloor during the earthquake had suddenly displaced a huge volume of the ocean's water. The effects rippled out in the form of a huge tsunami travelling at two-thirds the speed of sound. The term 'tsunami' comes from the Japanese for 'harbour wave'. While not always perceptible on the open ocean, the wave contorts into a mighty wall of water as the seabed rises to meet the land and can then devastate the shoreline.

As with earthquakes, Japan knows about tsunamis. The Pacific shore in the north of the country has had other large-scale tsunamis, including in 1611, 1896, and 1933. More recently, Akita Prefecture on the coast opposite from Sendai was inundated following an earthquake on 26th May 1983, becoming the most filmed and photographed tsunami up until that point. Living in an archipelago on a tectonic plate boundary forming part of the Pacific 'Ring of Fire' accustoms many Japanese to large waves from the sea and, just like earthquake warning systems, tsunami warning systems are commonplace throughout the country.

If the ground shaking in 2011 had not been warning enough to evacuate to high ground, then loudspeakers—those not knocked out by earthquake-related power failures—broadcast warning messages. The first tsunami wave reached the shores of Japan just fifteen minutes after the earthquake, although other places were not affected for an hour or two. Not everyone received messages, obeyed the warnings, or managed to reach places outside the low-lying zones. Cars floated as the water rose and then tumbled as the water writhed. People drowned in their houses, were caught in torrents racing along streets, or were hit by debris in the water. The tsunami's onshore height of over thirty-five metres in places, its inland reach of over fifteen kilometres, and its force, caught thousands off guard, including some professionals.

The emergency operations centre of Minamisanriku was located in a building three stories high. After the earthquake, the centre's workers leapt into action, doing their best to send out warnings about the imminent tsunami. Thousands of lives must have been saved by the information they disseminated. When the water advanced on their centre, the town's mayor stayed in the building along with dozens of employees, who ended up emailing their families to apologize for not leaving and who became increasingly alarmed as the ground floor of their centre flooded, followed by the floor above, and then finally the top floor. Two dozen people, including the mayor, reached the roof.

Photographed by Shinichi Sato, who was taking refuge atop a nearby building, the tsunami spilled over the centre's roof, with those on top clambering up the roof's antennas. As the tsunami's water and debris whirled around them, they clutched the antennas and each other, knowing that any slackening of grip, any succumbing to the current, any break of an antenna, would mean death. Over a dozen perished. The mayor and nine of his colleagues survived to descend the ravaged building, now little more than a metal skeleton, and make their way into a ruined town. The remnants of the building were left standing to serve as a memorial to those who died and as a tribute to the power of a tsunami.

Dramas similar to that played out at Minamisanriku were unfolding all along the coast, leading to an eventual death toll of over 15,000 people. The success in avoiding an earthquake disaster was superseded by a disaster resulting from the known, expected, but evidently in retrospect not adequately prepared for, hazard of a tsunami.

The following days and weeks had the world watching with bated breath. As the small coastal towns were looking for survivors and wondering if more waves would hit, the nuclear power plant of Fukushima, south of Sendai, was in trouble. Three of the six reactors were online during the earthquake and they shut down automatically, but the shaking cut out all external power to the plant. The tsunami inundated the Fukushima plant to a level of up to fourteen metres. Exact heights are estimates, because the deluge destroyed measuring gauges. These waves easily overtopped the four- to five-metre-high breakwaters (designed mainly to provide a safe harbour rather than for tsunami protection) and the ten-metre-high ground level upon which many of Fukushima's buildings sit. Water surged through the site, flooding buildings and stopping most internal power to the reactors, including the emergency generators. Pumps using seawater to cool the reactors could no longer function.

Just over twenty-four hours after the earthquake, when the workers thought that they might have the situation under control, one reactor exploded, releasing radioactive material into the air. Over the next few days, of the remaining five units in the power plant, two more exploded, one suffered major damage, and two which were being inspected at the time of the tsunami had little damage, partly because a single generator above the final waterline remained functioning.

The official report of the Japanese parliament spared no words in incriminating human choices for causing the disaster.[2] The government, regulators, and operators were condemned for not acting on their knowledge that a tsunami could inundate the Fukushima plant and that such an event would take out power with dire consequences. They had had numerous opportunities over the years to tackle the risk and chose not to do so, just as they chose not to prepare for the possibility of a major incident at the plant in which large amounts of radiation might be released. According to the report, all events on 11th March 2011 were foreseeable and preventable long before the earth shook and the waters moved. The Fukushima disaster was unequivocally unnatural.

Haiti's 2010 earthquake was over a hundred times less powerful than Japan's 2011 shaking, yet Haiti's earthquake-related death toll was over ten times the number of people Japan lost in the tsunami—remembering that few died in Japan's earthquake. The difference lay in the vulnerabilities of people and society, rather than in the hazards of earthquakes (and tsunamis), as causing the disaster and its impacts.

For the environmental events and processes we can deal with by reducing vulnerability, which are most of them, we are the real causes of disasters, not nature. Vulnerability arises primarily from the choices of those who have the power and resources to make those choices. Inadvertently or deliberately, in knowledge or in ignorance, disasters emerge through human choices, actions, behaviour, and values. We have achieved so much success in dealing with disasters and stopping human suffering, but we have a long way to go yet.

Closing this chasm between what we are achieving now to prevent disasters and what we should do is not easy. A debate surfaces. Some people aim to change fundamentals, focusing on the big picture in order to overcome baseline causes of vulnerabilities. In this book, we have seen examples of how discrimination, poverty, and incompetence feed into disasters. Others prefer to work on smaller scales and less ambitious steps. They demonstrate more direct, more tangible, and more immediate positive impacts which they hope, in the end, might scale up to wider, deeper changes. We have seen examples of such efforts here too, such as managing forests to permit small wildfires, not relying so much on flood walls, and retrofitting properties and surroundings—and changing our behaviour—to withstand wildfires and floods without harm.

One approach never precludes the other. We can practise both together, covering all scales of action in tandem so that they complement rather than impede each other. The argument for combining them is that the chronic human conditions of vulnerability which cause disasters are ever-present at all scales and must be tackled at all scales, especially over the long term, so disasters never appear rapidly. Rather than an event, we should recognize that a disaster is a long-term process.

Some hazards release their forces and energies swiftly with little specific warning. While we know broadly where earthquakes could strike at any time, such as Haiti and Jamaica, we cannot yet predict that an earthquake will occur in a specific place at a specific time. We know broadly where hurricanes could strike, also including Haiti and Jamaica, and we can observe the progress of a specific hurricane, but we cannot really say more than several days in advance when and where a major storm might make landfall. We know that Haiti and Jamaica are vulnerable to earthquakes, hurricanes, tsunamis, epidemics, and other hazards because of their social inequities and infrastructure inadequacies.

Poverty, ineffective governance, prejudice, and infrastructure lacking adequate planning and building codes all take a long time to become entrenched. Disasters, being caused by vulnerabilities, require the same lengthy time period to develop as these vulnerabilities. Consequently, in the same way that disasters are not natural, they are not unusual or extreme. They dramatically expose the vulnerabilities with which people live, and are typically forced by others to live, on a daily basis.

We can prevent 'natural disasters' and their human suffering, despite the presence of major environmental hazards, by reducing vulnerabilities. We must actively choose to do so.

NOTES

Chapter 1

1. United Nations, 2013. 'Statement of the Special Representative of the Secretary-General of the United Nations in Haiti A. I. Nigel Fisher, on the Compensation Case Filed Against the United Nations on Behalf of Victims of the Cholera Outbreak in Haiti'. MINUSTAH (Mission Des Nations Unies Pour La Stabilisation En Haïti), 21st February 2013. This statement begins: 'The United Nations informed the legal representatives of the claimants in the compensation case filed against the United Nations on behalf of victims of the cholera outbreak in Haiti, that their complaint is not receivable pursuant to Section 29 of the Convention on the Privileges and Immunities of the United Nations'.

2. United Nations, 2016. 'UN's Ban apologizes to people of Haiti, outlines new plan to fight cholera epidemic and help communities'. United Nations News, 1st December 2016. The Secretary-General stated: 'On behalf of the United Nations, I want to say very clearly: we apologise to the Haitian people. We simply did not do enough with regard to the cholera outbreak and its spread in Haiti. We are profoundly sorry for our role.'

3. H. Prichard, 1900. 'Through Haiti'. *The Geographical Journal*, vol. 16, no. 3, pp. 306–19.

4. J. Scherer, 1912. 'Great earthquakes in the Island of Haiti'. *Bulletin of the Seismological Society of America*, vol. 2, no. 3, pp. 161–80.

5. L. C. Ivers, 2017. 'Eliminating cholera transmission in Haiti'. *New England Journal of Medicine*, vol. 376, pp. 101–3.

6. K. Mika, 2019. *Disasters, Vulnerability, and Narratives: Writing Haiti's Futures*. Abingdon, Routledge.

Chapter 2

1. J. Leder, F. Wenzel, J. E. Daniell, and E. Gottschämmer, 2017. 'Loss of residential buildings in the event of a re-awakening of the Laacher

See Volcano (Germany)'. *Journal of Volcanology and Geothermal Research*, vol. 337, pp. 111–23.

2. R. Nash, 1985. 'Sorry, Bambi, but man must enter the forest: Perspectives on the old wilderness and the new'. Section 7, 'Banquet Address'. In J. E. Lotan, B. M. Kilgore, W. C. Fischer, and R.W. Mutch (technical coordinators), Proceedings—Symposium and workshop on wilderness fire, 15th to 18th November 1983, Missoula MT (pp. 264–8). General Technical Report INT-182 from US Department of Agriculture, Forest Service, Intermountain Forest and Range Experiment Station.

3. Parliament of Victoria, 2010. *2009 Victorian Bushfires Royal Commission: Final Report*. Parliament of Victoria, Melbourne, Victoria, Australia.

4. B. A. McCave, c.1928. *Captivating Canvey: An Isle of Delight in the Mouth of the Silvery Thames*. Canvey Island, Canvey Island Chamber of Commerce.

5. H. Bondi, 1967. *The London Flood Barrier: The Bondi Report and Branch Comments. Report to the Ministry of Housing and Local Government*. London, HMSO.

6. J. Jackson, 2001. 'Living with earthquakes: Know your faults'. *Journal of Earthquake Engineering*, vol. 5, supplement 001, pp. 5–123.

Chapter 3

1. K. Hewitt (ed.), 1983. *Interpretations of Calamity from the Viewpoint of Human Ecology*. Boston, MA, Allen & Unwin.

2. J. Lewis, 1999. *Development in Disaster-Prone Places: Studies in Vulnerability*. London, IT Publications.

3. Z. Qian, 2010. 'Without zoning: Urban development and land use controls in Houston'. *Cities*, vol. 27, no. 1, pp. 31–41.

4. JC Gaillard, K. Sanz, B. C. Balgos, S. N. M. Dalisay, A. Gorman-Murray, F. Smith, and V. Toelupe, 2017. 'Beyond men and women: A critical perspective on gender and disaster'. *Disasters*, vol. 41, no. 3, pp. 429–47.

5. Oxfam, 2005. 'Oxfam briefing note: The tsunami's impact on women'. Oxford, Oxfam International.

Chapter 4

1. E. M. Luna, 2003. 'Endogenous system of response to river flooding as a disaster subculture: A case study of Bula, Camarines Sur'. *Philippine Sociological Review*, vol. 51, pp. 135–53.

2. J. R. D. Cadag and JC Gaillard, 2012. 'Integrating knowledge and actions in disaster risk reduction: The contribution of participatory mapping'. *Area*, vol. 44, no. 1, pp. 100–9.

3. A. Kamradt-Scott, 2016. 'WHO's to blame? The World Health Organization and the 2014 Ebola outbreak in West Africa'. *Third World Quarterly*, vol. 37, no. 3, pp. 401–18. Also, L. O. Gostin and E. A. Friedman, 2014. 'Ebola: A crisis in global health leadership'. *The Lancet*, vol. 384, no. 9951, pp. 1323–5.

4. L. M. Stough and D. Kang, 2015. 'The Sendai framework for disaster risk reduction and persons with disabilities'. *International Journal of Disaster Risk Science*, vol. 6, no. 2, pp. 140–9.

5. P. Blaikie, T. Cannon, I. Davis, and B. Wisner, 1994. *At Risk: Natural Hazards, People's Vulnerability, and Disasters* (1st edn). London, Routledge. Followed by B. Wisner, P. Blaikie, T. Cannon, and I. Davis, 2004. *At Risk: Natural Hazards, People's Vulnerability, and Disasters* (2nd edn). London, Routledge.

Chapter 5

1. M. H. Glantz (ed.), 1994. *Drought Follows the Plow: Cultivating Marginal Areas*. Cambridge, Cambridge University Press.

2. J. G. Price, G. Johnson, C. M. Ballard, H. Armeno, I. Seeley, L.D. Goar, C. M. dePolo, and J. T. Hastings, 2009. *Estimated Losses from Earthquakes near Nevada Communities*. Nevada Bureau of Mines and Geology Open-File Report 09–8, Reno, Nevada.

3. F. Mulargia and A. Bizzarri, 2014. 'Anthropogenic triggering of large earthquakes'. *Scientific Reports*, vol. 4, article number 6100.

4. Intergovernmental Panel on Climate Change: https://www.ipcc.ch; and UN Climate Change: https://unfccc.int.

5. N. Schmidt (ed.), 2019. *Planetary Defense: Global Collaboration for Defending Earth from Asteroids and Comets*. Basel, Springer.

6. Y. Miura and Y. Noma, 1993. 'Identification of new meteorite, Mihonoseki (L), from broken fragments in Japan'. Lunar and Planetary Institute, Twenty-Fourth Lunar and Planetary Science Conference, Part 2, pp. 997–8.

7. A. Robinson and P. Talwani, 1983. 'Building damage at Charleston, South Carolina, associated with the 1886 earthquake'. *Bulletin of the Seismological Society of America*, vol. 73, no. 2, pp. 633–52.

Chapter 6

1. B. Ahmed, I. Kelman, H. K. Fehr, and M. Saha, 2016. 'Community resilience to cyclone disasters in coastal Bangladesh'. *Sustainability*, vol. 8, no. 8, article 805.
2. National Diet of Japan. 2012. *The Official Report of the Fukushima Nuclear Accident. Independent Investigation Commission*. Tokyo, National Diet of Japan.

FURTHER READING

Deloughrey, E., J. Didur, and A. Carrigan (eds), 2015. *Global Ecologies and the Environmental Humanities: Postcolonial Approaches*. Abingdon, Routledge.

Hewitt, K., 1997. *Regions of Risk: A Geographical Introduction to Disasters*. Essex, UK, Addison Wesley Longman.

Krüger, F., G. Bankoff, T. Cannon, and L. Schipper, 2015. Cultures and Disasters: Understanding Cultural Framings in Disaster Risk Reduction. Abingdon, Routledge.

McGuire, B., 2005. *Global Catastrophes: A Very Short Introduction*. Oxford, Oxford University Press.

Mileti, D. and 136 contributing authors, 1999. *Disasters by Design: A Reassessment of Natural Hazards in the United States*. Washington, DC, Joseph Henry Press.

Ripley, A., 2008. *The Unthinkable: Who Survives When Disaster Strikes—and Why*. New York, Crown Publishers.

Uitto, J. I. and R. Shaw, 2016. *Sustainable Development and Disaster Risk Reduction*. Tokyo, Springer.

INDEX

For the benefit of digital users, table entries that span two pages (e.g., 52–53) may, on occasion, appear on only one of those pages.